SUPERFUEL

SUPERFUEL

THORIUM, THE GREEN ENERGY SOURCE FOR THE FUTURE

RICHARD MARTIN

palgrave
macmillan

First published in 2012 by PALGRAVE MACMILLAN® in the United States—a
division of St. Martin's Press LLC, 175 Fifth Avenue, New York, NY 10010.

Where this book is distributed in the UK, Europe and the rest of the world, this
is by Palgrave Macmillan, a division of Macmillan Publishers Limited, registered
in England, company number 785998, of Houndmills, Basingstoke, Hampshire
RG21 6XS.

Palgrave Macmillan is the global academic imprint of the above companies and
has companies and representatives throughout the world.

Palgrave® and Macmillan® are registered trademarks in the United States, the
United Kingdom, Europe and other countries.

ISBN 978-0-230-11647-4

Library of Congress Cataloging-in-Publication Data

Martin, Richard, 1958–
SuperFuel : thorium, the green energy source for the future / Richard Martin.
    pages   cm
  Includes index.
  ISBN 978-0-230-11647-4 (hardback)
  1. Thorium. 2. Thorium—Separation. I. Title.
TP245.T6M37   2012
669'.2922—dc23
                                                                2011047557
669. 2922

A catalogue record of the book is available from the British Library.

Design by Letra Libre, Inc.

First edition: May 2012

10 9 8 7 6 5 4 3 2

Printed in the United States of America.

To Shawna, who listened and believed.

CONTENTS

Nature which governs the whole will soon change all things which thou seest, and out of their substance will make other things, and again other things from the substance of them, in order that the world may be ever new.

—*Marcus Aurelius,* Meditations

INTRODUCTION

One day in June 2009, I hiked along the Clinch River in northern Tennessee to what was once supposed to be the site of the most advanced nuclear power plant in the world. With my companions—John Kutsch, the owner of an engineering design firm in Chicago; Bruce Patton, a scientist at Oak Ridge National Laboratory 30 miles or so up the road; and Kirk Sorensen, an engineer at the Marshall Space Flight Center in Huntsville, Alabama—I clambered over a dilapidated chain-link fence and walked down the dirt road that followed the meandering river. It was the first really steamy day of summer. The woods were loud with crickets and desultory birdsong, and a double-crested cormorant launched itself off the surface of the sluggish river. We were trespassing on federal property, but it seemed unlikely that anyone would care. We passed the foundation of an old guard shack covered in foliage. It was like the setting for a postapocalyptic movie, except we weren't being pursued by zombies.

After a mile or so we came to a wide clearing on the inner curve of a horseshoe bend in the river. Obviously manmade, it was empty save for grass and gravel and a few arborvitae trees. Nothing stirred.

"Eight billion dollars," said Sorensen. "That's what you're looking at."

What we were looking at was the abandoned site of the Clinch River Breeder Reactor. Planned in the 1960s, Clinch River was originally conceived as the prototype of a new class of futuristic nuclear reactors that would create more fuel than they consumed. The project officially began in 1970 and finally was abandoned in 1983, after innumerable studies, reports, and rhetoric, plus the eight billion dollars that Sorensen mentioned. Once advertised as the future of power generation in the United States, Clinch River is now synonymous with technological hubris and the failed promise of atomic power. We were standing in the graveyard of the U.S. nuclear power industry.

As we ambled back toward our cars, Sorensen—who at that time was studying for a master's degree in nuclear engineering at the University of Tennessee—talked about the folly of U.S. nuclear policy and about the little-known element that could transform it.

"Thorium was the alternate path," he said. "It's a safer, more abundant fuel that could've revolutionized nuclear power. The problem is, it has almost nothing in common with what we're doing now."

I was on assignment for *Wired* magazine, and that trip, which encompassed a round of interviews at Oak Ridge and a driving tour of the Tennessee Valley Authority's extensive power-generation facilities, including the massive Watts Bar Nuclear Generating Station 25 miles or so to the south, was my introduction to the thorium power movement. This book is the outcome of the journey that started on that hot summer day in Tennessee.

I've written for *Wired* for ten years, and I've often joked that every story in the magazine has to use the word *revolutionary.* Thorium power actually seemed to justify that label. The more I learned about the lost history of thorium—especially the successful creation of a thorium-powered reactor at Oak Ridge in the 1960s and the career of the Oak Ridge director Alvin Weinberg, who championed safe thorium reactors and lost his job for it—the more astonished and outraged I became. Here was an inexpensive, safe, abundant energy source that could power every city on Earth, with enough left over for hundreds of millions of electric vehicles, for several millennia. And we were sitting on it, essentially doing nothing. It was insane. And the small band of technologists I was traveling with that day seemed like the only ones who were actually trying to bring the technology back to life, with zero encouragement from the government and plenty of disdain from the nuclear power industry.

TH90 • TH90 • TH90

IN A WAY THIS BOOK WAS BORN not when I first saw a mention of thorium power online (in a guest post written by Charles Barton Jr., on the website *The Oil Drum*) but decades before, when I was in high school, backpacking in the

Ozark Mountains of northwest Arkansas. I grew up in Little Rock, and my friends and I spent many, many weekends trekking through the Ozarks and camping in the thick forests that cover the deep gorges formed by the Buffalo River and its watershed.

For several years we went out every January 1 for a New Year's backpacking trip. The Ozarks are spectacular in the winter. Daytime temperatures fall into the teens, the creeks are frozen over so the hiking is easy (but not so solid that water's unavailable), and sunlight glitters through the bare elms, oaks, and walnuts. The solitude and stillness are absolute. Downed wood is plentiful in the Ozarks, and at night we built bonfires. Sitting around the fire, surrounded by the dark and frozen wilderness, forces on you a powerful impression of the energy that sustains our fragile lives. Backpacking requires you to carry with you all the fuel to sustain you for the length of the trek—for your camp stove, for your body, for igniting the energy stored in all that downed wood. Much time in the woods is spent calculating energy: How much should I eat now? How much wood do we need to gather? How much clothing should I wear to stay warm without sweating too much? How am I going to light my camp stove if my matches get damp? Will my flashlight batteries last?

I'm an early riser in the wilderness, and I used to awaken in the zero-degree dawn to light the morning fire. Relight it, actually: I prided myself on being able to restart the fire from the previous night's still-glowing embers, rather than using a match. That's when I began to wonder about energy. From where does it come? How is it that the glowing coals store their energy through the subfreezing night? Is energy infinite—could we ever burn all the dead wood this forest sheds? And so my thoughts ran until the sun came over the ridge and my companions emerged from their tents.

In a dim and intuitive way, I understood that energy knows no extinction, only transformation; that the energy in the wood, released as fire, is the same as the energy that fuels the stars that still throbbed in the lightening sky; that transformations of energy lie at the heart of all living things, and that energy lives in all things, no matter how dumb or inanimate.[1]

Later I pursued more defined answers through my work. I'd been covering energy for more than two decades, and it was increasingly hard to see a way out of our current fix. In 1987, during one of the periodic crashes that regularly decimated the oil industry back then, I spent a month driving the Gulf Coast, from Houston to the Florida Keys, to write about the people whose lives had been upended when the price of a barrel of oil went from $50 to $20 in a little more than a year. A decade later I was one of the first Western journalists in Central Asia to cover the Caspian Sea oil boom. I'd gone 600 feet underground to report on the world's richest uranium mine, at MacArthur River in northern Saskatchewan, and I'd flown across the frozen North Slope of Alaska to write about new horizontal drilling techniques that were pulling oil from under the permafrost near the Arctic National Wildlife Refuge. In 2001, a month after 9/11, I went to London to interview Royal Dutch Shell's team of futurists, the Scenarios Group. A couple of years later I was in the empty hull of an oil tanker in a New Orleans shipyard reporting on the plague of "super-rust" that was eating away the industry's supertankers. I'd come to respect, even admire, some of the people in the oil industry who I'd interviewed. But it was increasingly obvious that global climate change was going to wreak widespread havoc on, if not the complete destruction of, our oil-addicted society. I liked driving a car, and I liked being able to fly across the Pacific in a day. I didn't want to go back to a pre-fossil fuel economy.

For all the promise of renewable energy sources and all the hype lavished on them, it was clear that wind, solar, geothermal, biofuel, and such stood no chance of replacing a significant amount of carbon-based sources in time to significantly slow down the relentless heating of the planet. Only one source is clean enough, inexpensive enough, and abundant enough to do this: nuclear power. And despite the talk of a nuclear renaissance, nuclear power was going nowhere. Thorium has the potential to change that. After my *Wired* story on thorium ran in December 2009, thorium got little attention. An occasional newspaper feature or article in the science press would appear touting its promise, but no one else in the mainstream media was covering it in a systematic way. Here was a story that needed covering on an ongoing basis, and I was the only one on the thorium beat.

There was another factor at work. The thorium power movement was not static. In 2009 when I was reporting the *Wired* story, the thorium-heads, as I call them, were advocates, geeks, dreamers. They seemed like the kind of guys (they were all guys) who wrote letters to the editor of *Popular Science*. They had passion but no power, beliefs but no business plans. Their ideas were powerful, but they lacked the means to accomplish their desired ends. They discussed arcane topics in nuclear physics and materials science endlessly in online forums, conversations that never reached beyond the thorium community. As Kutsch, a pragmatist, liked to say, they were "building a boat in a basement."

Gradually that started to change. By the summer of 2011, when I was deep into the writing of this book, developments around thorium power were happening too fast for me to keep up. My story had played a small role in that—it helped Kirk Sorensen, who was featured in that story, get a job as the chief nuclear technologist at the nuclear supplier Teledyne Brown. Later he founded the thorium power start-up Flibe Energy. Groups like the Thorium Energy Alliance, a nonprofit formed by John Kutsch, moved the cause forward, too. Mainly, though, it was the involvement of actual business people (like George Langworth, featured in chapter 8), with actual business plans that had the potential to attract actual funding, that turned thorium power from a small movement of a dedicated few into a worldwide, if diffuse, program to build real reactors. Something big was afoot, and I was in position to witness it and tell the world about it.

TH90 • TH90 • TH90

ALL THIS WAS HAPPENING AGAINST THE BACKDROP of two wars in the Middle East, the rise of China as an economic superpower, total political paralysis in Washington, the worst financial crash since the Great Depression, and oil prices that looked to be headed permanently beyond the $100-a-barrel mark. In other words, most Americans did not have time to concern themselves with the disappearance of the Arctic ice cap or the inundation of South Pacific islands because of global warming. Few of the major accomplishments in American history have ever been achieved purely out of concern for the environment or for future generations. What I came to understand, though, is that thorium power is connected to just about all these issues and has the potential to solve, or at least address, many of the big problems we face as a nation and a society.

The development of safe, clean, essentially limitless thorium power technology could help right the U.S. trade imbalance with China. By supplying inexpensive electricity and desalinated water, thorium power could help cure some of the ills that still face the economies of the Middle East after the

Arab Spring. By creating jobs, calling forth a surge in technological innovation, and slashing the price of electricity, thorium power could help put the economy back on a sustainable track while providing opportunity and hope to millions of increasingly disenfranchised young people. It's a nonpartisan energy source that both Republicans and Democrats could get behind. Supplying clean, inexpensive electricity could help us wean ourselves from the ruinous dependence on imported oil. And so on.

It sounds like I believe that thorium is a panacea. And many in the energy industry will dismiss this book as a piece of misguided hype. The one question I hear more than any other when I describe thorium's potential is: "So why aren't we using it now?" The nuclear power establishment—the nuclearati, as I call them in this book—turns that question around as an argument against thorium: "If it was so great, we'd already be using it." This book, in a sense, is a long demonstration of why that's not so and why the structure of the nuclear power industry, its origins at the end of World War II, and its historical subservience to the interests of the military have prevented thorium from taking a place in the portfolio of world energy sources.

TH90 • TH90 • TH90

A FEW WORDS ABOUT THE TERMS "energy" and "power": Often used interchangeably, they're actually distinct. Put most simply, power is defined as energy expended or spent during a specific time interval. Think of a ton of wood. A certain amount of energy is stored in the wood. When the wood is burned, it releases energy in the form of heat. Energy is measured in joules; a ton of wood might have 18 billion joules of energy stored. If you burn that ton at a rate of one stick an hour, you'd release that energy at a certain rate, measured in joules per second—usually called watts. That rate is power. If you burn all the wood at once in a single conflagration, you get a higher level of power, but the total amount of energy doesn't change. Power is the rate at which energy is released.

Energy can be stored; power is an instantaneous measure. Energy can change its form; power cannot. To oversimplify, energy is the potential and power is the output. A thorium atom has a certain amount of energy stored within it. The rate of power output of a nuclear plant is usually measured in megawatts: a thousand-megawatt reactor, for example. The amount of energy actually released in a nuclear reactor is called the burnup. As I will show, the burnup achieved in thorium reactors is much higher than that of uranium. This is a major reason why thorium is such a better nuclear fuel than uranium.

I use "power" and "energy" in this book mostly according to their original meanings. The confusion of the two in the popular mind, though, leads to some misconceptions that affect our ideas about, and our policy toward, energy and power. We don't have an energy crisis. We have a power crisis. There is plenty of energy on Earth—streaming down from the sun, stored in deep geothermal reservoirs, packed into thorium atoms—to power human society effectively forever. The crisis lies in turning it into power—creating an output that can do work. Oil and coal, lying relatively close to the surface, easy to burn, with a high energy content by volume, were the easy sources to develop. With them we embarked on the Industrial Revolution, brought light and air-conditioning to millions of people around the world, fashioned a mobile car-based society, went to the moon and back, and built gleaming cities of skyscrapers where lights burn 24 hours a day. Now comes the hard part.

The choices we have to make are not easy ones. Thorium is no panacea, but of all the energy sources on Earth, it is the most abundant, most readily available, cleanest, and safest. We can't afford not to develop it. We also can't afford to continue making power choices based on politics, ignorance, pseudoscience, and the dominance of the rich over the poor. At its core this book is not just about a wondrous element that has the potential to solve our power crisis. It's about the choices we make as a society.

ONE

THE LOST BOOK OF THORIUM POWER

Kirk Sorensen was a rookie engineer at the Marshall Space Flight Center in Huntsville, Alabama, when he stumbled on the book that would change his life. This was in 2000. Sorensen was part of a team of engineers and physicists studying ways to use nuclear energy to power rockets to carry cargo into space. It was, as engineers like to say, a multivariable problem: the scientists had to consider the weight of the launch vehicle, tight confines of the engine compartment, extremes of temperature and atmospheric pressure as the rocket ascended beyond the atmosphere, risk of catastrophic accident, and so on. They quickly realized that conventional nuclear reactors would not do the job. And so they began looking into alternative reactor designs.

One afternoon that spring, Sorensen stopped by the office of his older colleague, Bruce Patton, a long-time nuclear engineer on assignment at the Marshall Center from Oak Ridge National Laboratory in Tennessee. Patton, who had lived through many changes of administration and many dead-end research programs at the national lab, had taken a liking to the young Mormon from Utah with a linebacker's build, a rocket scientist's intellect, and the temperament of a cattle-dog puppy.

Sorensen leaned against the door frame, his bulk filling the opening. The offices of chief scientists at Oak Ridge are not large, and Patton was not a chief. As technologists do, they chatted for a while in a language foreign to nonspecialists—Sorensen recalls it was about his growing frustration with the search for inexpensive ways to get heavy payloads into orbit. On the bookshelf in Patton's office he noticed a book with an intriguing title: *Fluid Fuel Reactors.* He picked it up and started leafing through it.

It was a book only an engineer could love. Published by the Atomic Energy Commission in 1958, during the Atoms for Peace era under President Dwight D. Eisenhower, and written by a group of contributors under the editorship of the Oak Ridge scientist James Lane, it ran 945 chart- and graph-crammed pages and weighed in at a biblical three pounds. Featuring chapter titles like "Integrity of Metals in Homogeneous Reactor Media" and "Chemical Aspects of Molten Fluoride Salt Reactor Fuels," *Fluid Fuel Reactors* details the work carried out in the 1950s at Oak Ridge, under then-director Alvin Weinberg. It describes nuclear power reactors with cores that were liquid, not solid, and that offered some intriguing advantages over the conventional light-water reactors (cooled by ordinary water) that make up nearly 90 percent of the reactors in operation today. It also describes the use of a novel nuclear fuel, an alternative to uranium and plutonium: the radioactive element thorium.

Sorensen took the book home and devoured it within days. His sleep suffered. A devout Mormon and a linear-thinking engineer, Kirk Sorensen was an unlikely revolutionary. But *Fluid Fuel Reactors* dropped a lit match into the dry tinder of his mind.

Here, he realized, was a potential solution—not to the problem of nuclear-powered spaceflight, which he had by that time decided was a pipe dream anyway, but to society's insatiable thirst for energy. Like most engineers of his generation, he knew that thorium is an actinide—one of the heavy elements on the bottom row of the periodic table of elements, a group that includes uranium and plutonium—and he vaguely remembered that the United States had done some work on thorium reactors in the two decades after World War II.

That work had gone far beyond calculations and experiments to an actual working reactor, and a sizable contingent of scientists, including Weinberg, believed that thorium-fueled reactors, with fluid cores of molten salt, should have been the future of nuclear energy. Outraged, Sorensen asked himself the question that has become a persistent refrain among thorium advocates: *Why has this never been pursued?*

Thorium is around four times as abundant as uranium and about as common as lead. Pick up a handful of soil at your local park or ball-field; it contains about 12 parts per million of thorium. The United States has about 440,000 tons of thorium reserves, according to the Nuclear Energy Agency; Australia has the world's largest resources, at about 539,000 tons. Like uranium and plutonium, thorium makes a dense and highly efficient energy source: scoop up a few ounces of sand on certain beaches on the coast of India, it's said, and you'll have enough thorium to power Mumbai for a year.

Used properly, thorium is also far safer and cleaner than uranium. Thorium's half-life, the time it takes for half of the atoms in any sample to disintegrate, is roughly 14.05 billion years, slightly more than the age of the universe; the half-life of uranium is 4.07 billion years. The longer the half-life, the lower the radioactivity and the lower the danger of exposure from radiation. Thorium's rate of decay is so slow that it can almost be considered stable; it's not *fissile* (able to sustain a nuclear chain reaction on its own), but it is *fertile*, meaning that it can be converted into a fissile isotope of uranium, U-233, through neutron capture, also known as "breeding." You can't mash together two lumps of thorium, even highly purified thorium, and trigger a nuclear explosion. Left alone, a chunk of thorium is no more harmful than a bar of soap. In fact, for a period before World War II, a thorium-laced toothpaste was marketed in Germany under the brand name "Doramad." Because of its unusually long decay process and its rare ability to breed through neutron capture, thorium is a more energy dense and efficient source of energy than uranium or plutonium: As a nuclear fuel, thorium reserves carry enough energy to power humanity's machines for many millennia into the future.

Thorium advocates point out that it's impossible to make a bomb from thorium, and significantly more difficult to make a bomb from uranium bred in thorium reactors than from enriched natural uranium. U-233 bred from thorium includes other undesirable isotopes, namely uranium–232, that provide built-in proliferation resistance. Nuclear waste from the thorium fuel cycle is also less hazardous to future generations. Fluid-fueled reactors known as liquid fluoride thorium reactors (LFTRs, pronounced *lifters*) can act as breeders, producing as much fuel as they consume. In LFTRs, thorium offers what nuclear reactor designers call higher burnup—there's less of it in terms of volume and less long-lived radioactive wastes to deal with afterward

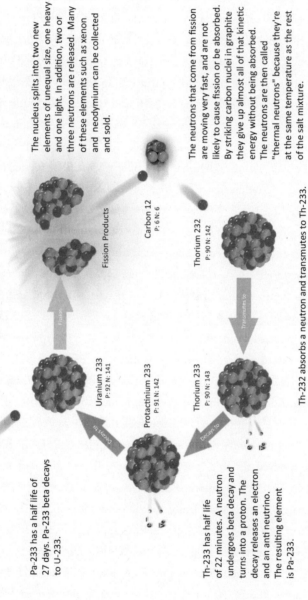

U-233 captures a neutron and fissions. When the atom fissions it generates 198 MeV of energy.

The nucleus splits into two new elements of unequal size, one heavy and one light. In addition, two or three neutrons are released. Many of these elements such as xenon and neodymium can be collected and sold.

Pa-233 has a half life of 27 days. Pa-233 beta decays to U-233.

Th-233 has half life of 22 minutes. A neutron undergoes beta decay and turns into a proton. The decay releases an electron and an anti neutrino. The resulting element is Pa-233.

Fission Products

Carbon 12
P: 6 N: 6

The neutrons that come from fission are moving very fast, and are not likely to cause fission or be absorbed. By striking carbon nuclei in graphite they give up almost all of that kinetic energy without being absorbed. The neutrons are then called "thermal neutrons" because they're at the same temperature as the rest of the salt mixture.

Uranium 233
P: 92 N: 141

Protactinium 233
P: 91 N: 142

Thorium 233
P: 90 N: 143

Thorium 232
P: 90 N: 142

Decays to

Decays to

Transmutes to

Fission

e⁻
ν̄e

e⁻
ν̄e

Th-232 absorbs a neutron and transmutes to Th-233.

In the thorium fuel cycle, thorium bombarded by thermal neutrons transmutes over a period of time to uranium-233, which is capable of sustaining a fission reaction. This process has multiple advantages over the fission process in a conventional reactor using uranium-235. (Brad Nielsen) than uranium. They can even consume highly enriched fissile material from dismantled warheads and long-lived transuranics in spent fuel from other reactors, turning it into a relatively benign and shorter-lived form of spent fuel, thus eliminating the need for geologic storage for thousands of years. What's more, LFTRs are inherently safe: The fission reactions occur in a radioactive cocktail of molten salt containing uranium-233 and jacketed by a blanket of thorium for breeding, requiring only a small start-up charge of enriched uranium, with thorium as the sole input thereafter. As the liquid fuel in the core heats up, it expands, decreasing the amount of fuel available, slowing the rate of fission reactions and cooling the fuel. It's like doubling the size of a pool table while keeping the number of balls on the table the same: fewer collisions occur, resulting in an extremely stable and responsive operation. The reactor core in a LFTR includes a "freeze plug" of frozen salt at the bottom, like the plug in a bathtub drain. Any power outage or unexpected deviation causes the freeze plug to melt, and allows the fuel in the core to drain into a shielded container designed to withstand the residual heat from the decay of fission products in the fuel. Because the reactor is inherently stable and the liquid fuel can be readily drained from the reactor core, a meltdown is physically impossible.

Thorium could provide a clean and effectively limitless source of power while allaying all public concerns—weapons proliferation, radioactive pollution, toxic waste, and fuel that is both costly and complicated to process. These concerns have crippled the nuclear power industry since the early 1980s.

Today, with global warming accelerating, climate-neutral nuclear power is poised for a worldwide comeback commonly referred to as the nuclear renaissance. At the same time, it's clear that the flaws of conventional, uranium-based nuclear power—which accounts for no more than one-fifth of power generation in the United States and less than that worldwide—make it an unsuitable replacement for fossil fuels in the near term. The nuclear accident that followed the earthquake and tsunami in Japan in March 2011 caused many countries to reconsider their ambitious nuclear agendas.

The problem is that only by shifting to non–carbon-emitting energy sources, like nuclear power, will we avoid catastrophic global climate change. Outside of the right wing of the Republican Party, hardly anyone today questions the worldwide scientific consensus that human-caused global warming, if left unchecked, will result in disruptions of a civilization-threatening nature: coastal cities like Calcutta and Miami inundated by seawater, huge swathes of farmland desertified, many now-populated areas uninhabitable, prolonged drought, and so on.

According to the International Energy Agency, worldwide demand for energy will rise by nearly 40 percent by 2035—a figure that many analysts, citing booming economic growth in the booming nations of China, India, and Brazil, consider low. Meeting that demand with current energy technologies would result in the addition of many billions of tons of carbon into Earth's atmosphere—and, most likely, in resource wars, famine, and the effective collapse of functioning society in many regions. The fossil fuel society on which we have built our civilization is simply no longer tenable.

Many well-meaning observers argue that by shifting to renewable sources, like wind and solar, and reducing energy demand through conservation and increased efficiency, we can shift away from fossil fuels in time to avert this disastrous scenario. Unfortunately, those hopes are illusory.

TH90 • TH90 • TH90

I FIRST MET KIRK SORENSEN in early 2009, when I was researching a feature for *Wired* magazine on the thorium power movement. I'd been covering energy for the better part of two decades. Since 2002 I'd been based in Boulder, Colorado, and had gotten a close-up view of both the natural gas boom that took hold in the northern Rockies in the first decade of the new century and the renewables push crystallized by Colorado's new governor, Democrat Bill Ritter. Like many of my generation, I had a deep foreboding about what rampant use of fossil fuels was doing to our planet and a conflicted attitude toward nuclear power.

I also had one overriding belief: A new technology that promises to improve life or provide people with new goods or make things less expensive cannot be stopped. You can delay it, regulate it, boycott it, or ban it, but eventually the technology will triumph.

I first read about thorium in a blog post by Charles Barton Jr., the son of one of the scientists who'd collaborated on experiments with thorium-based molten salt reactors in the 1960s. The blog ran on *The Oil Drum,* a "peak-oil" blog that examines the consequences of dwindling fossil fuel resources. At the same time, I was researching a report for Pike Research, a clean-tech energy research firm based in Boulder, on carbon capture and sequestration (CCS). CCS has been touted in some quarters as the answer to the evils of coal-fired power plants, which are by far the largest emitters of carbon per unit of power provided by any electricity source. Governments, including that of the United States, are pouring billions into developing systems that will separate the carbon from coal plant smokestack emissions (the capture) and then bury it permanently in underground reservoirs (sequestration). As I got deeper into the research, it became more and more clear that the numbers just didn't add up: CCS is an unproven, hugely expensive technology that is unlikely to be adopted at commercially significant levels in anything close to the time frames predicted by its supporters.

"Many of the current goals and targets for emissions captured between now and 2030 are overly optimistic," I wrote. This was a conclusion that was bolstered by studies from MIT, Stanford's Program for Energy and Sustainable Development, and Harvard's Belfer Center, among others.

Unfortunately, the same is true of many projections for renewable energy—especially solar and wind power. In the same period, I was doing some reporting for a website called *Energy Tribune,* run by Robert Bryce. A conservative journalist and energy analyst, Bryce has become one of the principal skeptics of green energy. I already had doubts about whether the glowing predictions for wind, solar, biofuels, and other forms of green energy could fulfill their promise in time to limit catastrophic global warming in my lifetime or that of my son, born in 1999. Bryce's work cemented those doubts.

In two books, *Gusher of Lies* (2008) and *Power Hungry* (2010), Bryce convincingly demonstrates that placing our faith in renewables, as that term is conventionally understood, is a "dangerous delusion." "The deluge of feel-good chatter about 'green' energy has bamboozled the American public and U.S. politicians into believing that we can easily quit using hydrocarbons and move on to something else that's cleaner, greener, and, in theory, cheaper."[1]

In fact, there is only one way to transition from an energy economy based largely on fossil fuels to a sustainable "New Energy Economy," as politicians like Colorado's Ritter like to call it: moving quickly to what Bryce calls N2N, a combination of natural gas and nuclear power for production of baseload electricity. (*Baseload* is the minimum amount of electricity that a power company must consistently generate to meet the demands of its business and residential customers.) In the liberal green circles in which I moved in Boulder, this amounted to right-wing heresy. But, while I didn't agree with some of Bryce's harder conclusions ("MYTH: Wind Power Reduces CO_2 Emissions"), the numbers were unassailable.

To take one example, the International Energy Agency has projected that new nuclear power plants will produce electricity for approximately $72 per megawatt-hour (one hour of operation at a rate of one megawatt). Electricity from onshore wind farms will cost up to $94 per megawatt-hour.

What's more, to build enough wind, solar, and other renewable energy projects to significantly reduce coal and oil use would require time and resources we simply do not have. In considering the real costs of different energy sources, it's important to take into account not just construction and operating costs but secondary factors like transmission, road building, resource extraction (of petroleum, coal, uranium, and so on), and real estate. Way back in the late 1970s, I took a course at Yale called "The Physics of Energy." The first assignment was to calculate how big a solar plant, in an ideal sun-drenched location like the American Southwest, would be required to supply 90 percent of U.S. electricity demand at the time. I'll spare you the calculations, but the answer was "roughly the size of the state of Arizona."

"Renewable sources such as wind and solar . . . require hundreds—or thousands—of square miles of land for power generation," Bryce noted. "The same problems of energy sprawl hamper the development of hydropower and biofuels."[2]

To give one more example, the local utility in Austin, Texas, where I spent a year in graduate school, announced in early 2009—just as I was becoming fascinated by the thorium movement—that it would spend $180 million on a 30-megawatt solar plant. Officials said the new sun farm would run at an average 23 percent of capacity, producing power at a construction cost of $6,000 per kilowatt of capacity. "Thus, Austin Energy has agreed to build a solar plant that will operate about one-fourth as often as a nuclear plant and cost about 25 percent more on a per-kilowatt basis," Bryce scoffed.[3]

And these power sources must be compared with uranium and thorium, the densest energy sources on the planet, which can produce power from room-sized reactors. Nuclear power came with its own costs, to be sure—costs that have been dismissed by the nuclear power industry—but, like Bryce and a growing number of green-energy supporters, I came to believe that wind and solar alone cannot help us escape from our current predicament. Only nuclear can. And only thorium can produce nuclear power that is environmentally safe, is economically competitive, and does not lead to the proliferation of atomic weapons.

For all those reasons, a critical mass of government funding, private-sector investment, and new research and development has begun to coalesce around the idea of returning thorium to its rightful place as a primary source of energy for the twenty-first century. Recently several countries, including the waking giants India and China, have announced or confirmed plans to build thorium power reactors, and R&D programs for thorium have sprung up at universities across Europe, Asia, and North America. Even U.S. policy makers have started to promote thorium power: several bills have been introduced to fund thorium R&D programs at the Department of Energy, and respected experts on climate change, such as the NASA scientist James Hansen, have spoken out in favor of thorium power. Thorium could transform not only the nuclear power industry but our entire energy economy, liberating us from dwindling oil supplies and poison-spewing coal plants. It could fuel the energy revolution the world desperately needs.

And Kirk Sorensen, in a modest engineer kind of way, has become the Lenin of this revolution.

TH90 • TH90 • TH90

WHEN I FIRST MET HIM, in the spring of 2009, Sorensen was a grad student in the University of Tennessee's storied nuclear engineering program and an obscure scientist on a NASA program that even its own researchers considered far-fetched. Two years later he was the chief nuclear technologist at Teledyne Brown Engineering, the aerospace, defense, and nuclear power arm of giant Teledyne Technologies, and he'd been officially assigned by Teledyne Brown's CEO, Rex Geveden, to build a thorium reactor.

In a small way I'd contributed to his rise. The story on thorium power I wrote for *Wired* magazine, which ran in December 2009 and featured Sorensen prominently, had made an impression on Geveden. "He said he read your story and in the margin he wrote 'Hire this guy,'" Sorensen told me later. By this time Sorensen had also become the unofficial general of the thorium power army, speaking frequently at conferences (including a couple of appearances at the Googleplex, the Silicon Valley headquarters of the search engine giant Google), and running the *Energy from Thorium* blog, which by early 2011 was getting about sixty thousand visitors a month. It is home to a discussion forum in which participants debate the finer points of nuclear arcana such Hastelloy-N, neutron flux, and the drawing-off of xenon gas. Sorensen was 35, and as I got to know him better I realized that the story of a man's life that pivots on the discovery of a book is central to his family's history.

He was born in Bountiful, Utah, and raised in Farmington, north of Salt Lake City, in the farmlands along the eastern shore of the Great Salt Lake. His ancestors, Danish Lutherans who converted to Mormonism, had settled the area more than a century before. Mormon missionaries had arrived in Denmark in the 1850s, and the new religion took hold in the minds of Kirk's great-great-grandfather, Isaac Sorensen, and Isaac's brother Abraham. Equipped with a copy of the Book of Mormon, the brothers convinced their extended family to join the young church and to make the long journey to Zion—that is, the unsettled frontier on the far side of the Rocky Mountains, in the New World.

"The time arrived for our leaving old Babylon, we took leave from the old homestead where father had dwelt 27 years and raised a family of twelve children," Isaac recalled, "ten alive, two dead."[4]

Approximately seventeen thousand people eventually left Denmark to settle in what became in 1896 the state of Utah. At one time Danish accents were heard among a significant proportion of the total Mormon population in the West. Most became farmers, but plenty also became builders. Isaac's great-grandson Keith, Kirk's father, spent his career working on big construction projects in Utah, including the massive pumps on the Great Salt Lake, and the Delta Center, where the Utah Jazz of the NBA play.

Kirk attended Utah State University (USU) to study mechanical engineering. Like most Mormons, he took a couple of years off from college to go on mission, proselytizing door to door like the missionaries who'd converted his great-great-grandfather. Sent to rough Texas towns like Palestine and Tyler, Kirk had less success. But Sorensen, a relentlessly glass-half-full guy, sees his missionary years as full of powerful lessons, if low on converts: "I got this incredibly concentrated life experience for two years—I had ten years of human contact in that time."

When he returned from his mission, Sorensen finished his studies at USU under a former U.S. Air Force colonel who impressed on his student the need to push forward with the exploration of space. With his customary alacrity, Sorensen moved into aerospace engineering. He quickly realized that the central problem of space exploration was the expense of launching vehicles into orbit: "If you couldn't lower the cost to put stuff in space, nothing else was going to matter." NASA was funding the X-33 program, a research effort to design a reusable launch vehicle that could reach orbit in a single stage. "I determined in the spring of '97 I was gonna get on this program," Sorensen said. After tracking down the email address of the chief engineer for the program, Sorensen wangled himself an internship on the X–33 team, headquartered at the Lockheed Martin R&D lab in Palmdale, California.

TH90 • TH90 • TH90

IT WAS A PIVOTAL POINT, a singularity, the first of several moments when his life took sudden right-angle turns. "Getting a job at Lockheed Martin was a bigger deal than my entire undergraduate education," he told me.

The next pivotal point came during graduate school at Georgia Tech. While he was working on a NASA-funded program led by John Mankins, the space agency's director of exploration systems research and technology, Sorensen was still studying reusable launch vehicles, now with the goal of putting massive solar arrays into space that would beam energy back to Earth by using microwaves. For an engineer Sorensen is a romantic. Like Conrad's Lord Jim, Sorensen had "that faculty of beholding at a hint the face of his desire and the shape of his dream." He read Stewart Brand's *Space Colonies*. He read *The High Frontier* by Gerald O'Neill. He became "a full-on 100-percent believer" that humanity's future lay in outer space. This was "the coolest crap in the world." It was utopia, and it was why aerospace engineers like Sorensen had to lower the costs of getting stuff into orbit.

Then, as I'd done with carbon capture, he tried to make the numbers work. He and his Georgia Tech colleagues spent months trying to "close the model," to work out a set of cost-benefit equations that would make space-based solar power economically feasible. How low would the cost of the rockets, and of the launch, have to be to make the business model realistic?

"We were continually trying to crank the launch cost lower and lower to get the model to close," Sorensen recalled. "We were modeling this incredibly dystopian future where a kilowatt-hour cost 25 cents, there was a carbon tax on everything, but we had perfect microwave beaming, and incredible solar arrays—every favorable factor for space-based solar."

Still, the numbers wouldn't crunch. "Finally I was sitting with my buddy and I said, 'Type zero into the launch model. How good is it at zero?' That meant, you snap your fingers and it's up there. He typed zero into the model and it still didn't close. We still couldn't make this thing work."

It was the first big disillusionment of Sorensen's young career. "I looked at him and said, 'Why are we working on this? This is a dodge. It's basically crap.'"

It was the end of his scientific innocence, of believing in a utopian energy future in which a career in space science was going to lead to a bright future for humanity. Sorensen's search, however, had only begun. He could not shake the idea that somewhere out there was a source of clean, inexpensive, limitless power, if only he could find the right technology. Next up was nuclear fusion.

At Georgia Tech he took a class with Weston Stacey, the author of several books on nuclear fusion and one of the many scientists who have basically spent their careers chasing the chimera of producing energy by crashing atoms together rather than splitting them apart. This time it took only a few months of study, rather than years, for Sorensen to conclude that he'd capture a live unicorn before he'd build a working fusion reactor.

"The more I learned, the more I was, like, 'You've got to be kidding—why on earth would anybody think this is going to make economic sense?'" He realized that, at its core, a fusion reactor relies on a giant superconducting magnet that literally wants to rip itself apart. In the old engineer's joke fusion is a technology that's 20 years off—and has been 20 years off for 50 years.

Like a man trying and rejecting belief systems, Sorensen was checking off a mental list of potential energy panaceas: Space solar, no. Nuclear fusion, nope. Studying fusion, though, led him to become intrigued with its more feasible cousin, fission.

This was at the beginning of the twenty-first century, and conventional fission-based nuclear power was seen, especially at places like Georgia Tech, as an old-school, unsafe technology that had had its moment and blown it. The nuclear industry in the United States was essentially dormant; no new reactors had been built for almost two decades, and few bright, ambitious technologists wanted to go into the field. Sorensen, though, became more and more intrigued. Fusion was difficult to the point of impossibility, he realized. Fission is not.

Often referred to as splitting the atom, fission is the process by which an atom absorbs a stray particle and then, after a fleeting instant, splits into a new atom and flings off various particles, releasing a tremendous amount of energy in the process. It's the transformation at the heart of the atomic bomb and of all nuclear reactors in operation today. Fusion works in the opposite way: two atoms combine, or fuse, releasing even more energy.

"The reason fusion is so hard is you've got charged particles, and they don't want to get anywhere near each other," Sorensen said. Overcoming those repulsive forces is incredibly difficult; just to get a tiny fraction to come close enough together to get a fusion reaction takes huge amounts of energy.

"Compare that with how easy fission is. A neutron doesn't feel the charge of the electrons that orbit the atom—it just waltzes through like there's nothing there. It doesn't involve high temperatures or thick magnetic fields or superconducting magnets or any of that," Sorensen said. "Fusion is so hard it takes teams of Ph.D.s and they're still scratching their heads. Fission is so easy we hire high school grads to run nuclear submarines and they can do it."

Sorensen, though, was an aerospace engineer by training, and he and his wife had a newborn. Switching careers at that point was hardly an option. So despite his misgivings about the future of space exploration, he took a job at NASA. Once again he found himself working long days to find inexpensive ways to get stuff into space. That's when he walked into Bruce Patton's office and found a copy of *Fluid Fuel Reactors* on the shelf.

TH90 • TH90 • TH90

AS HE READ THE BOOK, Sorensen quickly knew that he'd reached another pivotal point, and the subsequent course of his life seems to him now to have been as surely determined as his great-great-grandfather's was when he first held in his hand a copy of the Book of Mormon. Sorensen, though, claims the path was less straight. He'd been burned twice already by ideal-sounding solutions to the interlaced problems of insatiable energy demand and rising global temperatures; he would not be won over a third time quite so easily.

"I started thinking a lot about energy—what on earth are we going to do? When I started to learn about thorium, and about liquid fluoride reactors, I wondered, Why aren't people doing this? But then I was very skeptical—I'd had two energy lovers already."

First, he realized that, as an aerospace engineer reading a nuclear engineering text, he didn't understand the terminology. So he began to teach himself the relevant physics and engineering, reading the most important texts—many of which were hopelessly out of date by the early twenty-first century and hadn't been revised in decades—and following the discussions on websites like *Atomic Insights*. He read everything by Alvin Weinberg he could find, including the former Oak Ridge director's 1994 memoir, *The First Nuclear Era: The Life & Times of a Technological Fixer*. As Sorensen sought to understand the debates about the looming energy crisis, he studied the politics and economics of climate change. And he came to realize that the full history of thorium power, and of fluid-fueled reactors, lay moldering in boxes in lab buildings at Oak Ridge.

By this time, early 2002, Sorensen and Patton had been assigned to another task, one that seemed even more far-fetched than space-based solar arrays: researching the feasibility of a manned spacecraft to Callisto, one of the moons of Jupiter. To get there at the speed of the fastest known spacecraft would take years and way too much fuel for the rocket to carry. The mission would clearly require some form of nuclear power. Thinking back to Weinberg's work, Sorensen recalled the molten salt reactor experiment at Oak Ridge National Laboratory.

The Pentagon created Oak Ridge in 1942 in the hills outside Knoxville, Tennessee, and it served as one of three primary R&D centers for the Manhattan Project. It was, in many ways, the birthplace of both nuclear weapons and nuclear power. But the lab never did a perfect job of preserving its own history. Most of the history of the work done on the thorium fuel cycle and liquid-core reactors resided in dusty boxed archives, in reports and studies that had lain untouched since they were first produced and filed away. When Sorensen embarked on his quest to learn everything he could about liquid-core reactors, it was a rare moment when one man's intellectual inclinations joined perfectly with the requirements of a large bureaucracy. Sorensen and Patton managed to slice $10,000 from the research funding for the Jupiter mission study ($150,000 or so) to pay to have the documents at Oak Ridge on molten salt reactors (MSRs) exhumed, dusted off, and scanned into digital form. At the time, Sorensen had no intention of starting a movement; he was just doing a thorough research job. Having the material on CD would be much, much easier than combing through all that paper.

The documents, most bound in heavy reports, were digitized during the next six months. No more money was available. Sorensen now estimates he has about two-thirds of the most interesting material from the archive. It shows a far greater depth of R&D on MSRs in the 1960s than Sorensen, or indeed most of the current generation of Oak Ridge scientists, realized. Sorensen began to understand what a treasure he'd unearthed. Then, a few months later, history crashed down around the Marshall Space Flight Center.

On February 1, 2003, the space shuttle *Columbia* exploded on reentry, leaving a swath of shattered debris across east Texas. The entire space program was instantly thrown into a disarray from which it has never recovered; the idea of a manned mission to Jupiter suddenly seemed not just preposterous but insane. Along with a bunch of other NASA engineers, Sorensen began, out of necessity, to contemplate his next career move. He had already begun to think of turning away from outer space and delving into the smallest objects in the universe. A supervisor at NASA asked him, "Why don't you get a graduate degree in nuclear engineering?"

At the time this hardly seemed like a practical suggestion. The nuclear industry had been in the doldrums since the early 1980s, and Sorensen already had one master's degree. "I remembered graduate school," he said. "It was painful. I was poor and hungry." Needless to say, that didn't stop him.

The nuclear engineering program at the University of Tennessee (UT), which grew out of the lab at nearby Oak Ridge, had not only a storied history but an extensive distance-learning program; Sorensen could stay in Huntsville, keep his job at Marshall (such as it was), and take a class a semester. And the head of the department, Harold L. "Lee" Dodds, had co-authored an important paper on MSRs with Uri Gat and Dick Engel, two of the researchers who worked at Oak Ridge on the MSR experiment under Alvin Weinberg. In the fall of 2003, hardly a propitious time for nuclear power, he started at UT.

Although by that time the so-called nuclear renaissance had begun to gather momentum, nuclear power remained a backwater for promising young scientists and engineers. The near-disaster of Three Mile Island in 1979, and the full-blown disaster at Chernobyl seven years later, which together brought the nuclear power industry to a near standstill in the United States and many European countries, had made nuclear engineering as a profession not only unpromising but also uncool, sinister, and vaguely fascist. Most of the men (they are almost all male) working in the industry in the first decade of the twenty-first century were middle aged and beyond, and few younger engineers were moving up through the ranks to replace them. At one point, around the turn of the century, graduate programs related to nuclear power in the United States, the country that invented nuclear power technology, had fewer than 200 graduate students. Gradually this has changed in the ten years since; nuclear power is now seen as a promising field, in part because lots of people in their sixties and seventies are retiring, and no one in their thirties and forties is available to replace them.

Kirk Sorensen was aware of this gap, but it wasn't the motivating force behind his career shift. He wanted to change the world, and he'd become convinced that nuclear power—in the form of MSRs, or, more specifically, thorium-based liquid fluoride reactors—was the way to do it. But he quickly ran up against the inherent conservatism of the industry he wanted to transform. Sorensen's new professors at UT, including Lee Dodds, basically said, "If thorium power was so great, it would've been done a long time ago." It would never happen in their, or Sorensen's, lifetime.

That exchange well sums up the attitude of what I've come to call the nuclearati—the tight fraternity of utility executives, university physicists, nuclear engineers, policy makers, and investors who have, for 30 years, constituted the leadership of the nuclear power industry. In the wake of the Fukushima-Daiichi nuclear accident in Japan, it would be easy to scoff at men like Dodds. These are the same guys, after all, who told us for three decades that nuclear power would be safe and inexpensive. But Dodds is not just the prisoner of his own outworn assumptions; he is representative of a class of men who have spent their careers believing in, and building an industry around, nuclear power. They built the only significant worldwide energy source not based on burning some form of carbon, and they did it in less than a quarter century. It was a remarkable achievement, and it shouldn't be discounted because the uranium atom has proven less tractable, and the shifting breezes of politics and public perception less predictable than once believed.

Sorensen, at any rate, was once again undeterred. His belief in thorium was undimmed. And as he made his way slowly through the mysteries of nuclear physics and quantum mechanics, he began to think of taking on more of a public advocacy role in the emerging debates about the future of energy.

So he did what any passionate technophile with a cause to promote would have done: he started a blog. Called *Energy from Thorium,* it debuted on April 22, 2006. The Iraq War was at its bloody peak; the price of oil, which had lingered below $25 a barrel for nearly 25 years, had climbed to almost $60 on its way to nearly $135 in mid-2008. With his poll numbers nosediving, President George W. Bush, a former oilman, had signed an energy-efficiency bill on January 1. Global temperatures were on their way to another of the hottest years on record (according to NASA's Goddard Institute, nine of the ten hottest years have come since 2000).

Energy from Thorium "is intended to be a location for discussion and education about the value of thorium as a future energy source," Sorensen wrote in his debut blog post. "Despite the fact that our world is desperately searching for new sources of energy, the value of thorium is not well-understood, even in the 'nuclear engineering' community."

During the next several months, he posted lengthy articles on the history of liquid fluoride reactors and the differences between thermal-spectrum and fast-spectrum reactors (briefly, while the latter operates at high energies, with nothing to moderate, or slow down, emitted neutrons, the former uses a moderator, like water or graphite, to slow down neutrons in order to promote more fission reactions; see chapter 3). He asked, and answered, the question, "How much thorium would it take to power the whole world?" (Answer: about 1,500 metric tons, or something like 2 percent of the annual uranium consumption of the world's nuclear power industry today.) He also began posting the Oak Ridge MSR documents that had so painstakingly been digitized and burned onto CDs. (When I asked him if that material was in the public domain—and thus not subject to copyright protection—he answered, "I figured they were. Nobody ever said otherwise. I never asked Oak Ridge. I did ask if they wanted a copy of the CDs, but they didn't even want that.")

At first writing his blog was like dropping stones into an empty lake. But traffic to the site started climbing steadily, and gradually Sorensen realized that others had experienced the same nuclear epiphany he had. The early proponents of thorium tended to be middle-aged men, mostly with technical backgrounds, many of them lifelong tinkerers and cogitators who shared deep misgivings about the world their grandchildren were going to inherit and a passionate belief that nuclear power, properly developed and safeguarded, could supply the world with the clean, abundant energy it needed. They also had in common a vague disgruntlement: frustrated at the reduced influence of science and engineering in national policy, they suspected that the world undervalued the expertise they'd spent lifetimes accruing. In this, again, they resembled the nuclear establishment. But they were convinced that conventional uranium-based reactors, operated by a hidebound and risk-averse nuclear power industry, were not the answer. Thorium was.

Robert Hargraves had been a math professor at Dartmouth, a consultant for Arthur D. Little for 12 years, and the chief information officer of Boston Scientific, the medical device company. He remembered thorium from his doctoral studies in physics in the 1960s, and when he began to see the element mentioned again in energy discussions, he decided, in retirement, to devote his remaining years (and a chunk of his savings) to it. A frequent commenter on *Energy from Thorium* and other nuclear blogs, Hargraves taught a class called Rethinking Nuclear Power at Dartmouth's Institute for Lifelong Education. Joe Bonometti, a West Point grad and a former professor in space systems at the Naval Postgraduate School, worked on the same NASA team researching nuclear-powered spacecraft as Sorensen and Bruce Patton. Like them, he became convinced that thorium in liquid-core reactors represented a fundamental breakthrough that had been too long ignored by the nuclear power industry. Bonometti began leading some of the informal R&D initiatives that arose from the discussion forum attached to *Energy from Thorium*, aiming to solve some of the outstanding engineering issues surrounding the forgotten technology.

Bonometti's goal, like that of the nascent thorium community as a whole, was "to ensure that no one in government, or industry, or anywhere else can claim they were blindsided or somehow secretly held in the dark about the technology."[5] They didn't want to just promote a neglected energy source; they wanted to save the world with liquid fluoride thorium reactors. Technology had gotten us into this mess. The thorium-heads believed that technology could also save us.

John Kutsch ran a design engineering firm in Chicago. A wealthy customer who'd made a fortune in real estate asked him in 2005 to look for investment opportunities around various minerals, including thorium. Kutsch found his way to *Energy from Thorium* and began communicating with Kirk Sorensen. Like many people new to the thorium discussion, Kutsch was floored to learn that "we have this resource and we're doing nothing with it," as he told me in 2009. Kutsch, who bears a certain resemblance to George Costanza from *Seinfeld* but with more hair, had made an unsuccessful run for Congress a few years earlier. He quickly realized that while Sorensen, Patton, Bonometti, and the rest of the nascent thorium underground had a deep understanding of the technological and scientific issues, they had little capacity or inclination to get things done in the real world.

"The folks at the *Energy from Thorium* forum are the best there is for technical answers," Kutsch told me, "but I'm afraid that spending time debating the best way to prevent xenon taint and what to do with transuranics et cetera won't get anybody any funding, or really anywhere else, any time soon."

In other words, you can't cross the ocean by debating the best sailboat designs. You have to build a boat. Determined to do something pragmatic and effective to move the energy industry toward a thorium-based nuclear future, Kutsch set up the Thorium Energy Alliance (TEA), a nonprofit that has organized a series of conferences on thorium power. The TEA managed to

catch the attention of the deepest-pocketed player in the new-energy field—
Google, which had committed $1 billion to funding renewable, carbon-free
energy that is less expensive than coal. In the spring of 2010, I spoke at a
conference at Google headquarters in Silicon Valley that was hosted by Dan
Reicher, the former assistant secretary of energy who headed up Google's
clean-energy arm. (Reicher has since left Google to direct the Steyer-Taylor
Center for Energy Policy and Finance at Stanford.) By early 2011, Kutsch
had become one of the most visible activists in the thorium underground,
spending much of his time lobbying Congress and raising money—while
trying not to neglect his own business.

Then there was Charles Barton Jr. The blog post by Charles Barton on
the peak-oil blog *The Oil Drum* was what first alerted me to the thorium
movement, and Barton was the first thorium advocate I spoke to in January
2009. He had inherited his obsession with alternatives to uranium-fueled
light-water reactors. His father Charles Barton Sr. was a chemist who'd gone
to work at Oak Ridge in 1948, at the dawn of nuclear power, to investigate
the chemistry of elements for use in power reactors. Along with many of his
colleagues during Alvin Weinberg's tenure as director, Charles Sr. became
swept up in the MSR program—a forerunner of today's LFTR (explained
in more detail in chapter 6)—and wound up becoming the world's foremost
authority on the chemistry of heated liquid salts. Like other Oak Ridge scientists, Charles Sr., who died in January 2009, had spent the twilight of his
career watching his work get dismantled and his research discarded. His son,
a drug and alcohol counselor, made it his life's mission to see his father's work
vindicated. Thorium reactors, the younger Barton wrote, will "open up a
source of carbon-free energy that can last centuries, even millennia."

LFTRs, Barton told me, could be built far smaller and less expensively
than conventional light-water reactors: "You could build a whole bunch of
small reactors, truck them out to sites around the country where you need
power plants, dig holes in the ground, put them in, and turn them on."

TH90 • TH90 • TH90

IT IS, OF COURSE, NOT THAT SIMPLE. I came to realize fairly soon that the tone of
the *Energy from Thorium* forum—geeky, high minded, theoretical, and naive—characterized the thorium movement as a whole. It seemed clear that a

small band group of advocates, however committed, had little chance of influencing national energy policy or turning the giant battleship of the nuclear industry.

"The nuclear industry has zero incentive to shift to a new fuel cycle," Charlie Hess told me. A long-time executive at the architectural engineering firm Burns & Roe, Hess spent 30 years building and operating nuclear plants. Although he is a prototypical member of the nuclearati, he is an advocate of alternative nuclear power, including thorium-based reactors, and a critic of the nuke-power establishment. Fuel costs for uranium reactors are less than half a cent per kilowatt-hour. "They spend more on security guards than they do on fuel," Hess told me. "Frankly they don't care."

That was made clear to me by John Rowe, the CEO of Exelon, the country's number one producer of nuclear power, when I pulled him aside after a speech at a National Press Club luncheon in Washington, DC. When I asked about the possibility of shifting to thorium as a primary nuclear fuel, he assured me that there "will be alternatives across the entire fuel cycle." But inexpensive uranium works just fine for Exelon, which has a market capitalization (the total value of its outstanding shares) of $28 billion and made $18.6 billion in revenue in 2010. If it's not broke, don't fix it—and nuclear tycoons like John Rowe have convinced themselves that the nuclear power industry is not broken. From the perspective of his office suite, that's certainly true: Rowe made $10.3 million in 2010, and between 2006 and 2011, his compensation totaled $153.9 million. Uranium reactors have been good to nuclear power executives.

Rowe's dismissive attitude embodies the obstacles that face the thorium movement, which is composed of outsiders. "Look, the nuclear industry in the U.S. is very conservative," Ambassador Thomas Graham told me. "I can see interest here in the U.S. gradually developing. But it's not going to happen here first." Graham, a long-time diplomat and opponent of nuclear proliferation who served as a President Bill Clinton's special representative for arms control, now chairs the board of Lightbridge, a company based in McLean, Virginia, that is developing solid fuel thorium rods for conventional reactors. While Graham foresees the use of thorium in the American nuclear power industry at some point, "the initial deployments," he said, "are going to be abroad."

Abroad. In the three years I've been covering the thorium movement, almost every conversation has at some point included that stipulation. The United States, which dropped the first atomic bomb on Japan at the conclusion of World War II, pioneered nuclear power, built the first commercial power reactors, and invented the liquid-core reactor and first proved that thorium could be used in power-generating reactors, is, barring some unforeseen and unlikely shift in energy policy, almost certainly destined to be a laggard in the worldwide thorium revolution.

France is the world's largest producer of nuclear power and supplier of uranium for reactors. Eighty percent of its electricity comes from nuclear power, and the energy giant Areva has an active thorium R&D program and is investigating the possibility of building liquid fluoride thorium reactors by 2032. The Laboratoire de Physique Subatomique et de Cosmologie in Grenoble is the only facility in the world that has the resources and backing needed to actually develop a commercial LFTR by 2022.

The Řež nuclear research institute in the Czech Republic is a leader in the development of MSRs and is investigating the possibility of fueling MSRs with thorium, according to the institute's director.[6] Norway, which has an estimated 180,000 tons of thorium reserves, is embarking on an ambitious long-term nuclear power program that includes the construction of thorium-fueled reactors. In Brazil, which has the world's second-largest thorium reserves and began research into thorium power in the 1960s, R&D efforts have recently begun again to develop thorium-fueled solid fuel reactors. By far the most active thorium power programs, however, are in Asia—particularly in the emerging economic superpowers of India and China.

In February 2011, China officially announced that it will start a program to develop a thorium-fueled molten salt nuclear reactor, taking a crucial step toward replacing coal with nuclear power as a primary energy source. The program was announced at the annual conference in Shanghai of the Chinese Academy of Sciences and is headed by Jiang Mianheng, son of the former Chinese president Jiang Zemin and the holder of a Ph.D. in electrical engineering from Drexel University. The People's Republic has no intention of falling behind in the race for the next great energy source.

The world's most ambitious thorium power program, though, is in India, which has the world's largest thorium reserves. India exploded its first nuclear weapon in 1974 in defiance of the Nuclear Nonproliferation Treaty, and it has always viewed nuclear energy—in both warheads and power reactors—as a key element of national sovereignty. The country has embarked on a three-phase program to build as many as 60 reactors, converting them to run on thorium before 2032.

I will detail the Indian and Chinese programs in chapter 7 and the implications for the United States in the conclusion. Here it is enough to quote the 2011 film *The Ides of March,* in which the progressive presidential candidate, played by George Clooney, declares, "Either we're going to lead the world or we're going to bury our heads in the sand." The question of thorium is not whether it will become a major source of energy—it will—but when—and where and who will lead the way.

TWO

THE THUNDER ELEMENT

Thorium is a lustrous silvery-white metal, denser than lead, that occurs in great abundance in Earth's crust. It's slightly radioactive, but, like naturally occurring uranium, you could carry a lump of it in your pocket without harm. When heated, thorium incandesces a brilliant white. Its atomic number—the number of protons in an atom's nucleus—is 90. On the periodic table it's found on the bottom row, along with the other heavy radioactive elements, or actinides—protactinium, uranium, neptunium, plutonium, and so on. With an atomic weight of 232, it is the second-heaviest element found in measurable amounts in nature, behind uranium (atomic weight essentially equals the number of protons plus the number of neutrons). It has a half-life—the time it takes for half of any sample to decay to a nonradioactive state—of about 14 billion years, or about the age of the universe.[1]

Thorium is around four times as abundant as uranium, or about as prevalent as lead. The international Nuclear Energy Agency estimates the United States has 440,000 tons of thorium reserves.*

It was discovered in 1828 by Swedish chemist Jöns Jacob Berzelius, who named it for the Norse god of thunder. Berzelius, who was born in 1779, originally trained as a physician, and his early medical studies included some

---

* The Nuclear Energy Agency is a division of the Paris-based Organization for Economic Cooperation and Development.

daring ideas, among them the influence of a galvanic electrical current on various diseases. More practically, he helped develop the technique of electrolysis, or using direct current to stimulate a chemical reaction, and invented the method of notating chemical formulae that is still in use today. In his career he also discovered selenium and cesium, and he was the first to isolate a long list of elements that includes calcium, barium, silicon, and titanium. Along with Robert Boyle and Antoine Lavoisier (the French chemist who devised the first periodic table and whose law of the conservation of matter would be overturned by Einstein's theory of relativity), Berzelius is considered one of the fathers of modern chemistry. On thorium, though, he had already made an embarrassing blunder: he'd mistakenly "discovered" it once before.

People who knew of Berzelius's fame, and his skill at identifying strange new materials, had a habit of bringing him samples to test in his lab at Stockholm. In 1815 he obtained an unfamiliar material, a black earth, and subjected it to the usual chemical analysis. At the time he believed he had discovered a new element. Wishing to honor the Scandinavian deities, he named it after Thor. As Berzelius's knowledge of chemical interactions deepened, however, he began to have doubts about thorium, and in 1824, almost a decade after he'd announced his discovery, he realized that the black earth was actually a form of a previously discovered element called yttrium.

Four years later he got another chance. A Swedish pastor named H. M. T. Esmark was hiking on the island of Lövö, off the west coast of Norway, when he spotted a black mineral with a darkly gleaming surface. The islands of Norway are formed of granite and gneiss, with stony uplands dotted by fantastic outcroppings. Reverend Esmark reportedly liked to walk these desolate heaths, glorying in the works of the Lord and occasionally collecting interesting rock samples for his father, a prominent mineralogist named Jens Esmark.

In this case Professor Esmark could not identify the material. So he sent it on to Berzelius, who at age 49, despite the thorium misstep, was considered the leading chemist in Sweden. This time Berzelius was certain he was right, and in 1829, after subjecting the material to a series of chemical analyses and isolating the 60 percent of the sample that was an unknown element in pure form, he announced the rediscovery of thorium in a paper in a Swedish geo-logical journal. It was 40 years after the discovery of uranium by a German apothecary and 66 years before the accidental discovery of radiation by the German physicist Wilhelm Conrad Röntgen.

Like all heavy elements, the material Berzelius identified is literally not of this earth. All the thorium on Earth was created in supernovas—the tremen-dous stellar explosions that mark the end of large stars' stable life.

Nothing is lost in the immolation of a star. In the core the elements hy-drogen and helium are combined in nuclear fusion, in which smaller elements collide and fuse to make larger ones—including the basic stuff of life, carbon, nitrogen, and oxygen. Stellar fusion, however, is not powerful enough to cre-ate elements heavier than iron (atomic number 26). In the cataclysm of star death, the heart of the star heats a furnace in which new heavy elements are formed, a process called nucleosynthesis. These new elements are flung into the vacuum of deep space like embers from a conflagration. Debris condenses in the gravity of other, younger stars to form rocky or gaseous planets. Our solar system took form from the vast elemental clouds drifting in space and came into being roughly 4.5 billion years ago. At Earth's center thorium and uranium give off the energy from long-dead stars.

That energy, in fact, still fuels the heat generated underground in the mantle that underlies the continent-forming crust. Earth generates an enor-mous amount of radioactive heat—something on the order of 2.1×10^{13} watts, or about 20 million megawatts. Heat emanating from the depths cre-ates Earth's magnetic field (which shields the planet from the corrosive solar wind), enables the plate tectonics that formed today's continents, and sus-tains life itself. Only three radioactive elements are found in enough abun-dance to generate significant amounts of heat in the terrestrial skin: uranium, potassium, and thorium. Three isotopes of uranium occur naturally: U-238, U-234, and U-235. The vast majority of the uranium on Earth—about 99.3 percent—is U-238. For use in conventional reactors uranium must be en-riched to 3 to 5 percent U-235, a blend known as low-enriched uranium (LEU). Highly enriched uranium, anything better than 20 percent, can be used in nuclear weapons, although for a reliable modern warhead guaranteed to produce an explosion equivalent to 465 kilotons or so of TNT and to destroy a good-sized city, you need uranium that is about 85 percent pure

U-235. The hard part about building an atomic bomb is not building the bomb; it's the expensive and complicated process of enriching enough uranium to cause an uncontrollable fission reaction.

Like potassium, thorium has only one naturally occurring isotope, Th-232, with 90 protons and 142 neutrons. (The difference between U-235, with 143 neutrons, and thorium, with one fewer neutron, is a critical one.) The decay of these three elements is responsible for the great majority of geothermal heat; uranium, with a half-life that is 10 billion years shorter than thorium's, produces more heat by volume than thorium, but thorium is much more abundant in Earth's crust. It is safe to say that if not for the thorium created in the explosion of dying stars five or six billion years ago, life on Earth would not exist, and you would not be reading this book.

Like all radioactive elements, thorium has a characteristic decay chain—the series of elements into which the material spontaneously transforms as it sheds particles in the form of radiation. The decay chain of thorium includes two isotopes each of radium and polonium; another thorium isotope (Th-228); radon; and, eventually, lead, the stable resting element into which both uranium and thorium decay. Almost all the thorium currently in Earth has been present since the planet was formed.

A fissile element is one whose atoms will split apart when bombarded by low-energy neutrons. The primary fissile elements are uranium-235, plutonium-239, and uranium-233. Naturally occurring uranium-238 is not fissile; that's why you have to enrich it to increase the percentage of U-235. Thorium is not fissile, but it is fertile: under the right conditions—say, bombarded by neutrons in the core of a nuclear reactor—it can be converted into U-233, a highly fissile isotope of uranium. Thorium-232 actually captures a neutron to become Th-233, which then decays quickly, by way of protactinium-233, into U-233. The difference between fissile U-235 and fertile thorium plays a large role in this story.

Although it was critical to the late nineteenth- and early twentieth-century physicists who blazed the pathway into the mysteries of the atom, thorium is much less well known than its cousins uranium and plutonium. The dramatic part played by uranium in the history of the twentieth and the twenty-first century completely overshadows thorium's potential and its unique powers. If elements were celebrities, thorium wouldn't even make the B list.

In many ways thorium is a shadow element to its more infamous neighbor on the periodic table, uranium: chemically similar and exhibiting closely related but utterly tangential behavior, the two are deeply linked on some elemental level. Thorium is the first element into which U-238 decays; in a nuclear reactor thorium transmutes into an isotope of uranium, U-233, that has qualities more suited for power generation than the U-235 version. It is as if thorium had a complementary or contrasting feature for every response and quality of uranium. Yin and yang. Masculine and feminine. Light—the light of Trinity, of Bikini Atoll, and of Hiroshima—and shadow.

Thorium could be the younger sister—less volatile, slower to self-consume—of her flamboyant and domineering older brother, uranium. Their differences have defined the history of nuclear power—and on the threshold of World War II this pair of radioactive siblings enabled some of the key theoretical insights that led directly to the development of nuclear weapons.

TH90 • TH90 • TH90

"IT IS REMARKABLE THAT THE TWO most active elements, uranium and thorium, are the ones which possess the greatest atomic weight."

In that short sentence, included in the epochal 1898 paper, "Rays Emitted by Compounds of Uranium and of Thorium," Marie Curie embedded many of the concepts that would underlie that next half century of astonishingly rapid discoveries in nuclear physics, quantum mechanics, and the science of nuclear fission.

From Henri Becquerel's accidental discovery of radioactivity in 1896 to August 6, 1945, when Hiroshima was destroyed by an atomic bomb, occurred without question the most remarkable series of discoveries—*revelations* is a better word—in the history of science. This quest of modern physics was catalyzed in 1905 by Albert Einstein's insights into the inextricable identity of energy and matter; pushed forward by a remarkably colorful and brilliant group of European physicists who were led, driven, and inspired by the Danish master Niels Bohr; and culminated, in the high desert of New Mexico, in the realization of the worst fears surrounding the power of the shattered atom. Because the quest culminated in the utter destruction of two ancient cities—not military targets—by the most fearsome weapons conceived up to that point, it is a story of absolute betrayal, of the subversion of humanity's highest ideals and its most brilliant minds by a technology that led irrevocably to a fundamental change not only in warfare but the human condition and potentially to the eradication of the human race itself.

This is the fundamental dilemma of twentieth-century nuclear physics: this band of brilliant, humane, cosmopolitan scientists led us all to the brink of annihilation. In *Faust in Copenhagen,* his entertaining account of the famous Copenhagen gathering of 1932, the physicist Gino Segre makes it clear that by the mid-1930s, as fascism spread malignantly across Europe, Niels Bohr, Werner Heisenberg, Paul Dirac, Leo Szilard, Enrico Fermi, and the rest possessed a dawning awareness of the awful trade-offs—the Faustian bargains—they were making on the road to full understanding of the power that lurked inside radioactive elements. I'd read that story in several well-told and exhaustive accounts, most powerfully in Richard Rhodes's *The Making of the Atomic Bomb.* I considered myself, as a science-and-technology journalist, fairly well versed in the subject. Now, though, while researching thorium's tale, I realized that there was a shadow version, and, seen through the lens of thorium, it cast the whole enterprise in a new and surprising light. A year after I first heard of thorium, I came to see it as a key element in the remarkable history of discoveries that transformed our world in the first half of the twentieth century.

My purpose here is to examine how the story of thorium is woven throughout the astonishing scientific and technological advances of that magical and terrible half century—to shed some light on a shadowy element.

In the case of Marie and Pierre Curie, the distinctions between uranium and thorium opened the way to their greatest discoveries. Since Becquerel's discovery of radioactivity, Marie Curie—whose biography is a prefeminist tale of unthwarted ambition and conventional barriers overcome—had become fascinated by this effect. The mystery of Becquerel's "rays," she would recall later, "seemed to us very attractive and all the more so because the question was entirely new and nothing yet had been written upon it."[2] What's more, the radioactive properties of thorium were still undiscovered; Berzelius, working before radiation was even detectable, had no idea that the element he discovered was one of the most radioactive in nature.

In 1894, having completed her math degree at the Sorbonne, Marie Sklodowska had set up a laboratory in a disused storage space, little more than a closet, in the school in Paris where Pierre Curie (whom she would marry a year later) taught. As a woman and a Pole, Marie had endured terrible hardship during her early years as a student in Paris, but she quickly set her challenging circumstances to productive use. Along with Pierre, she "arrived at 'a new method of chemical analysis' based on very precise measurements of what we now call radiation."[3] She started out with uranium, which Becquerel had used, but she quickly moved on to other elements, including thorium.

To test the properties of these materials she used leftover wooden grocery crates to build a primitive but extremely sensitive electrometer, which had been devised by Pierre and his brother. Inside were two metal plates held one above the other. The lower plate, charged with a high-voltage battery, held the material being tested; the upper one could be tested for electrical current. Sklodowska, a gifted scavenger, first used white uranium powder obtained from the French chemist Henri Moissan, who had supplied Becquerel with the same material. After measuring the current produced in the upper plate by uranium, she moved on to other elements, including gold and copper and a bizarre substance called pitchblende. Mined in Joachimsthal in eastern Germany, a major silver-mining district since the sixteenth century, pitchblende was a black tarry mineral from which Martin Heinrich Klaproth had first extracted uranium in the late eighteenth century. Then she tested the mineral aeschynite, which she knew contained thorium.

The literature and Marie's detailed lab diary are not clear about where Curie obtained her thorium. Some of her mineral samples came from chemist friends of Pierre's; some came from the Museum of Natural History in Paris. Wagons regularly arrived outside the school to deliver loads of exotic minerals. At any rate, the results of the examination were startling.

"I have studied the conductance of air under the influence of the uranium rays discovered by M. Becquerel," she wrote in the 1898 paper, "and I examined whether substances other than compounds of uranium were able to make the air a conductor of electricity."[4] The other substances included various oxides of uranium and thorium as well as potassium, sodium, ammonium, chalcite, and pitchblende.

At this point, at the end of the nineteenth century, uranium was seen as a curiosity with few practical applications beyond its use as a coloring agent for ceramics. Thorium, on the other hand, was already the key element of a thriving industry, mostly centered on the manufacture of mantles for gas lanterns. Readily available, thorium was prized above other metals for its capacity to phosphoresce brilliantly at high temperatures. First invented in the 1700s, gas lighting became a major source of illumination in the early nineteenth century. By the 1830s much of Paris was lit by gas streetlights, earning the French capital its sobriquet "City of Light." None of the strollers on the Boulevard Haussman in Belle Époque Paris realized that the streetlights contained one of the greatest sources of energy on Earth, one that helped power the fires at the core of the planet.

In fact, Marie Curie was not the first to discover that thorium is radioactive: that honor technically goes to Gerhard Carl Schmidt, a German chemist who reported the finding in the *Deutsche Physikalische Gesellschaft* in March 1898, a few weeks before Curie presented her paper to the French Academy of Sciences. Schmidt, however, delved no further into thorium's mysterious properties; he promptly disappears from the story.

Marie Curie's 1898 paper gave the magnitude of the current, in amperes, emitted by the substances she tested using Pierre's electrometer and a piezoelectric quartz (*piezoelectric* refers to the ability of certain quartz crystals to transform kinetic energy into electrical current; tap one end with a hammer and a minute electrical current emerges from the other). To her surprise, the most active material was not purified uranium; it was raw pitchblende, followed by thorium oxide. Thorium, in fact, was nearly twice as potent as uranium (pitchblende was nearly four times as strong).

The astonished scientist realized that the ability to produce an electrical current in air was not unique to uranium; it was a fundamental property of the universe, found to varying degrees in many elements. Not only that, but the qualities of this strange activity could be used to discover new and even stranger elements.

With this understanding the revolution began. Marie Curie, though, had noticed another unique aspect of thorium, one that she did not pursue: the current produced by thorium oxide was not steady. Early on Curie realized that the activity in the electrometer would gradually increase when she was testing thorium. "She was intrigued enough to open the chamber up, renew the air, and make further measurements. Once again, she noticed a slight increase in activity within the chamber."[5]

"Had these experiments been more clear-cut," wrote Irene Curie, Marie's daughter, many years later, "the entire orientation of future work might have been changed."[6]

For the first of many times, thorium's shadowy qualities were pushed aside in favor of uranium, which afforded more clearcut results. In a few more years the question of thorium's unstable activity would arise again.

TH90 • TH90 • TH90

WHILE MARIE CURIE WAS TESTING SUBSTANCES in her wooden electrometer, a young New Zealander working at the Cavendish Laboratory at Cambridge under J. J. Thompson, the discoverer of the electron, was rapidly turning over the tables of fusty British physics. The New Zealander was Ernest Rutherford (later the first Baron Rutherford of Nelson), a peerless deviser of experiments who worked with a series of gifted assistants. He essentially fathered twentieth-century experimental physics, discovering the nucleus, chronicling the transformation of elements, and revealing the power of the free neutron, all during 25 years of groundbreaking work. Born in 1871, he made many of his most famous discoveries after winning the Nobel Prize in Chemistry in 1908. In 1898, the year of Marie Curie's discovery of radiation, Rutherford joined the faculty of McGill University in Canada—not exactly a hotbed of advanced physics research at the time—where he would carry out a famous series of experiments in the following decade that would transform the understanding of matter and lay the groundwork for the discovery of nuclear fission in the interwar years.

The Curies had declared that the mysterious source of high-powered radiation inside Joachimsthal ore was an entirely new element, which they christened radium. Although its dreadful effects would not become apparent for years, the therapeutic wonders of radium were understood almost immediately; used to shrink tumors and to effect various other, less tangible cures, it quickly became the focus of a burgeoning medical industry that brought not only fame to the Curies but also wealth, almost all of which they plowed back into the lab. After thorium, radium became the second radioactive element to generate a lucrative commercial industry.

Much about these invisible rays was still not understood, and Rutherford, along with an enterprising young British graduate student named

Frederick Soddy, performed a series of ingenious and influential experiments to tease out their nature. Rutherford had already noticed the same effect that Marie Curie had: thorium's activity seemed to fluctuate over time. In 1899 one of Rutherford's McGill assistants confirmed that the radiation from thorium changed "in response to events as irrelevant as the opening of a door." Apparently thorium was sensitive to "slight currents of air." Intrigued by this "capricious variation" of thorium, Rutherford performed a series of experiments to study it. In the summer of 1899 Rutherford solved the puzzle "and in so doing shook the foundations of chemistry as it was then known."[7]

He found that thorium emits a gaseous substance that is different from the thorium itself. Rutherford had discovered the alchemist's dream: the transmutation of elements, only not in response to some magical solution but simply as part of the strange processes at the heart of the atom. He called the new substance an "emanation" and reported that it has the power "of passing through the thin layers of metals, and, with great ease, through considerable thicknesses of paper." The radioactivity of the new substance decreased in a geometric progression: half of its original strength after one minute, half of that half in another minute, and so on. What's more, the emanation (which Rutherford suspected was a gas) "possesses the power of producing radioactivity in all substances on which it falls," and the new, induced radiation "is of a more penetrating character than that given out by thorium or uranium." Rutherford called this new activity "excited radioactivity."[8]

At almost the same moment, the Curies, their imaginations fired by Rutherford's relentless experimentation, were also uncovering evidence of excited radioactivity caused by radium and polonium. Rutherford, enlisting the help of Soddy, continued his work on the thorium emanation. Rutherford noted that "the duration of the conductivity" in ionized gas in his testing apparatus would have been far less if the experiment had been performed on a compound of uranium. Thus the first distinctions between these two closely similar elements began to emerge.[9]

But Rutherford was after bigger fish. Surmising that thorium's emanation consisted of some kind of radioactive gas, he set Soddy—fresh from Oxford, a talented chemist, and an experimentalist almost equal to Rutherford himself—to work discovering its identity. Through a series of painstaking chemical analyses—physics was far more laborious in the days before research reactors and particle accelerators—Soddy came to "the tremendous and inevitable conclusion that the element thorium was slowly and spontaneously transmuting itself into argon gas!"[10]

The emanation was in fact an inert gas incapable of reacting or combining with any other substance. In the fall of 1901 Rutherford and Soddy came to the conclusion that had been inescapable all along: the thorium was disintegrating. One element was being transformed into another. "Standing there transfixed as though stunned by the colossal import of the thing," Soddy later recalled, he blurted out, "'Rutherford, this is transmutation!'"[11]

The two scientists fired off a paper to the *Journal of the Chemical Society* in London; in the paper they stated that radioactivity is "at once an atomic phenomenon and the accompaniment of a chemical change in which new kinds of matter are produced." It was "one of the major discoveries of twentieth century physics," Rhodes declares, and in it rested the destruction of Hiroshima and Nagasaki, the dawning of the Atomic Age, and the promise of thorium power.[12]

Working with thorium and other radioactive substances, Rutherford and Soddy in quick succession calculated the half-life of uranium and thorium and their decay products, defined the concept of an isotope (the term was Soddy's), and distinguished between different forms of radiation: beta radiation comprises highly charged electrons, while alpha particles are actually helium atoms ejected during radioactive decay. After providing the emanations that enabled Rutherford to shatter the assumed indivisibility of the elements, thorium, as is its wont, sank from view, only to be taken up again a few decades later. In the interval I must pause to explain in more detail the nature of the atoms that Rutherford and his contemporaries were probing.

TH90 • TH90 • TH90

IMAGINE A HARD PLASTIC SPHERE like a child's ball. Now imagine a cluster of plastic balls, some colored red, some blue, all held together by an invisible but unfathomably strong glue. Other spheres (colored black, let's say) circle this cluster in distant orbits and are held on their courses by some mysterious force, also powerful but many orders of magnitude weaker than the glue that holds the center cluster together.

This is the conventional representation of the atom, familiar to us from ten thousand logos. Like most such representations—including Niels Bohr's early conception of the atom as a "liquid drop" with a smooth surface, it is inaccurate. So strange is the world of fundamental matter that it is literally impossible to picture it faithfully in our familiar physical terms. For one thing, the black spheres—electrons, discovered by Rutherford's mentor J. J. Thomson in 1897—do not really orbit. (In fact, under the principles of quantum mechanics,* it is impossible to say exactly what they do.) For the purposes of an explanation of nuclear fission, though, it's the best available image.

The red spheres are protons. The blue are neutrons; the black, as I have said, are electrons. Electrons are negatively charged, protons positive. Neutrons, as another Rutherford protégé, James Chadwick, demonstrated in 1932, have no charge at all. This is the key that would unlock the awesome energy at the heart of matter.

At the end of the nineteenth century, when scientists first began to understand that certain forms of matter emitted rays, or tiny fragments named particles, they knew that the positively charged protons (red) would necessarily attract the negatively charged electrons (black). Neither particle could move freely through intermatter space. There had to be a third particle, with "zero nuclear charge" (as Rutherford put it in a 1920 lecture to the Royal Society in London), that "should enter readily the structure of atoms, and may unite with the nucleus." He would later describe the mysterious particle as "an invisible man passing through Piccadilly Circus—his path can be traced only by the people he has pushed aside."[13] This was the neutron, the blue ball, which can pass easily through the electron barrier to shatter or, in some cases, impregnate the nuclei of other atoms.

Some atoms, as I have said, decay spontaneously. In the process they give off three types of radiation: alpha (which consists of essentially helium ions, or the nuclei of helium atoms), beta (energetic electrons), or gamma (photons, with zero mass and no electric charge). Some are more prone to disintegrate than others when struck by neutrons. In the plastic ball model it is as if the nuclei of some elements—the heavy, radioactive ones on the bottom rows of the

---

* Werner Heisenberg's Uncertainty Principle states that it is impossible to accurately measure both the location and the momentum of a particle.

periodic table—are too densely packed, too loosely held together, to be stable. Balls fly off spontaneously as the cluster of spheres shifts to another form; some of the blue balls (neutrons) strike other clusters, disturbing them, and more balls fly outward in three-dimensional space. A certain number collide with more clusters and so on. If more than one blue ball (neutron) is ejected from every nucleus-struck ball, the reaction will continue almost ad infinitum.

When a stray neutron penetrates another atom's nucleus, two things can happen: the atom splits (fission), or the atom absorbs the neutron to become a new isotope. I will explain further in chapter 3 how fission is controlled, and the speeding neutrons moderated, in a nuclear reactor; here I focus on the phenomenon itself. Experiments in harnessing fusion to drive generators and make electricity have been going on since the 1960s, to little avail. This was the dismal record that drove Kirk Sorensen to place his faith in the future of fission. So it is fission that I am explaining.

When the red and blue balls (protons and neutrons) split apart, they break the invisible glue (known as the strong force) and release incredible amounts of energy. Even before the full implications of fission were understood, physicists from Einstein onward realized how awesome that energy is. A single gram of matter contains 85 million British thermal units of heat, enough to generate 25 million kilowatt-hours of electricity. The trick to nuclear power is making that energy available without blowing everything to smithereens. Under the right conditions, the atoms of certain forms of matter—the fissile elements uranium and plutonium—can be induced to disintegrate in a predictable and orderly fashion. Because atoms are so tiny, the amount of energy released per collision is small—200 million electron-volts (MeV) or so—but there are many, many collisions, and their number increases exponentially if the reaction is uncontrolled. Unlike uranium, all of the thorium in a given volume can be "burned"—i.e., converted into fissile U-233 to power a reactor, meaning that thorium requires no expensive enrichment to be used as a nuclear fuel. Nobel laureate Carlo Rubbia estimates that one metric ton of thorium can produce as much energy as 200 metric tons of uranium.

It might appear, at first, that the faster and more energetic the neutrons are, the more effective they'll be in shattering other atoms, but that's not the case. Slower neutrons have a higher probability of interacting with other atoms. Physics instructors often use the baseball metaphor: the slower the pitch, the better chance the batter has of connecting with it. (In fact, conventional uranium reactors require a moderator—often graphite or plain water—to slow down the neutrons so as to increase the probability of further fission reactions enough to sustain the chain reaction in the core. See chapter 3 for more on how reactors work.) Here it's worth keeping in mind the dimensions of the stadium around the batter.

One of the most astonishing discoveries of early twentieth-century physics was the vastness of inner space. As with the nonexistent orbits of the electrons, the metaphor of the solar system is often used to convey the gulfs of distance between the nucleus (the sun in this model) and the electrons (the planets). Another way to think of it is a large city plaza, say, St. Peter's Square. If the atom were the size of the square, the nucleus would be the size of a grain of sand.

Put another way, a cubic meter of lead contains less than 1 percent actual matter, in the form of subatomic particles. The rest is vacuum. In the plastic ball model, the black balls are very, very far from the tiny cluster of red and blue balls at the center.

Nevertheless, the chances of one neutron smashing into another atom, in a common material like thorium-232 or uranium-238, are quite high. But the chances of fission happening, as opposed to absorption, are relatively low. That's why naturally occurring uranium decays, but the ore does not fission: a much denser material—a critical mass—or a higher percentage of U-235 is required to produce the explosive chain reaction of a nuclear warhead. And thorium won't chain-react at all, unless it transmutes into U-233. An image favored by nuclear engineers is a room filled with hundreds of mouse traps baited with ping-pong balls. When one trap goes off, the ping-pong ball flies off and hits another, setting it off, which in turn releases a ball that springs another trap. If the traps are packed closely enough, a single one sprung sets off a chain reaction until almost all go off. If they're too far apart, as in natural uranium, the reactions quickly peter out.

By the last few years of the 1930s, as fascism swept across central Europe and war became inevitable, most of this had been understood and documented. Physicists had understood that new, transuranic (beyond uranium) elements could be created by bombarding heavy elements with neutrons. They knew that certain elements, including thorium and uranium, spontaneously decayed. At the end of 1938, thanks to the work of the radiochemist

Otto Hahn, they realized that a neutron collision can overcome the tremendous binding force of the nucleus and split an atom apart—a process that the Austrian physicist Otto Frisch dubbed fission. Only one mystery remained, the solution of which would lead directly to Hiroshima, the nuclear arms race, and the dawn of nuclear power: Could fission be induced and prolonged to create a self-sustaining nuclear reaction?

The answer, as with other breakthroughs in this period of astonishingly rapid advances, would come through close examination of the related-yet-opposing qualities of uranium and its sister element, thorium.

TH90 • TH90 • TH90

ON A VISIT TO THE UNITED STATES in early 1939, Niels Bohr had a flash of insight that would change the understanding of fission and the power of radioactive elements. Over breakfast one morning at the Nassau Club at Princeton, Bohr was discussing his liquid drop model of the atom with a skeptical Polish physicist named George Placzek.

Placzek pointed out a fundamental inconsistency in the prevailing understanding of the fission characteristics of uranium and thorium: while both tended to fission when bombarded by high-energy, fast neutrons (of more than one MeV), only uranium atoms split under bombardment by slow neutrons. Thorium, more resistant than uranium, was impervious to slow neutrons. "If the liquid-drop model had any validity at all, the difference made no sense," Placzek argued.[14]

One of Bohr's outstanding talents was the ability to shift course when presented with evidence that contradicted his notions. By this time he was less an active experimentalist than a teacher and leader, a mentor to younger scientists, and a guide and supporter of promising new research. But he was still capable of remarkable leaps of intuition. That day he had another such leap. He and Placzek hurried to Bohr's office.

"Now listen," Bohr exclaimed. "I have it all."

He drew a series of graphs with the horizontal axis showing neutron energy, left to right, and the vertical axis depicting the cross section of nuclear reactions—the probability of a certain reaction's occurring, increasing as the graph climbed. The affinity of nuclei to neutrons of a certain energy is called resonance. Both thorium and uranium are resonant with neutrons at more or less the same levels of energy, capturing them readily. Bohr was plotting the resonance curve of the two elements, or how they responded to neutrons of varying energies. He drew one graph for thorium, one for uranium-238, and one for uranium-235, the lighter and more rare isotope. The first two graphs were identical; the one for U-235 was quite different.

Up until this point, physicists had assumed that the differences in the fission profiles of thorium and uranium were more or less inconsequential. While each had its specific advantages for certain types of experiments, the differences had been considered an artifact of no great import. Placzek's questions reexamined that assumption, and Bohr's graphs disproved it. By comparing the resonance curves of thorium and uranium, he established that inside uranium there lurked a demon, an isotope that would burst apart upon encountering any neutron, of any energy. Tickle it and it would explode. Bohr understood that this tiny fraction of natural uranium was what makes a chain reaction—and thus a bomb—theoretically possible. Bohr rushed out a paper in early 1939.[15] The invasion of Poland was less than seven months away. World War II had not yet begun, but Bohr had just opened the last gates to the scientific breakthrough that would decide it. Separating out the U-235, however, was fiendishly difficult, so much so that "it would take the entire efforts of a country to make a bomb," Bohr stated. He was, of course, right.[16]

TH90 • TH90 • TH90

THE FINAL STOP ON THIS THORIUM-BASED TOUR of nuclear physics history is Berkeley, California, where in the late 1930s Glenn Seaborg found and described the transuranic elements. The transuranics are elements with atomic numbers higher than 92. None occurs at more than trace levels in nature. By this time nuclear testing equipment had progressed far beyond Pierre Curie's crude electrometer. Working on the Lawrence Cyclotron at the University of California at Berkeley, Seaborg and his graduate student assistant were able to bombard any element of their choosing with neutrons and analyze the results with exquisite precision.[17]

The career of Glenn Seaborg, one of the most remarkable scientist-administrators ever produced by the United States, exhibits the Janus-faced quality of many of the pioneers of nuclear physics: his work looked in one direction to developments that would vastly improve people's lives and gazed in the other at the ultimate destruction of human society by nuclear weapons. The discoverer or codiscoverer of ten elements, he was a pioneer in nuclear medicine, the use of radioactive isotopes to diagnose and treat disease. He was the discoverer of plutonium, and he found uranium-233, the fissile isotope into which thorium-232 transforms and that has the potential to solve our current energy crisis and fuel a new era of inexpensive carbon-free energy.

After getting his Ph.D. at Berkeley in 1937, Seaborg set out in the footsteps of Frederic Soddy to uncover and investigate different isotopes of radioactive elements. At that time scientists understood that transuranic elements were not only possible but also, according to the tenets of theoretical physics, had to exist. There had to be an element 93 and an element 94, but no one had managed to produce them in the lab or to understand their properties. With his colleague John Livingood, in 1938 Seaborg had already created a new isotope of iodine, iodine–131, which is still used today to treat thyroid disease. Now he was hunting a very different beast.

In 1940 Seaborg was part of the team led by Edwin McMillan that discovered element 93, which they named neptunium. Then McMillan was called away to work on the development of radar technology for the war effort and granted his protégé permission to take the next step: the search for element 94.

Seaborg had perfected the oxidation-and-reduction technique, conceived by McMillan, that had tracked down neptunium. Now he and his colleagues used the same method to show that a new element could be formed from the induced decay of neptunium. In February 1941, after bombarding a sample of uranium, they isolated plutonium-239, the long-sought ninety-fourth element. A month later Seaborg showed that plutonium, like U-235, is highly fissile. Because separating U-235 from U-238 was such a difficult and costly process, the realization that plutonium, produced through the relatively simple bombardment of uranium with neutrons, was fissile led directly to the Manhattan Project and to the plutonium bomb that destroyed Nagasaki.

Seaborg understood the gravity of his discovery. He spent much of the rest of his life working for arms control. More immediately he turned to thorium.

Seaborg was curious whether thorium could be transformed into a fissile isotope. He knew of another isotope of uranium, U-233, which is found in trace amounts in nature but is not part of the direct decay chain of natural U-238. Suspecting that it might arise from thorium, he instructed his grad students to turn the cyclotron on thorium-232.

First they produced thorium-233, which had a half-life of only 23 minutes, decaying to protactinium, element 91. Little was known about protactinium at that time, other than that it had properties similar to zirconium (atomic number 40) and a half-life of about 27 days. Logically, because the protactinium decayed with the emission of beta radiation (an electron), the next step on the decay chain had to be one unit higher on the periodic table: uranium-233. The sequence looks like this:

$$Th\text{-}232 \rightarrow Th\text{-}233 \rightarrow Pa\text{-}233 \rightarrow U\text{-}233$$

And, sure enough, after bombarding a large amount of thorium, letting the protactinium decay, and concentrating the result, Seaborg and his team showed that they had created U-233, effectively a new isotope of uranium. When they tested it under bombardment, they discovered that not only was it fissile, but it produced more than two neutrons per fission reaction.

Though the full implications of his discovery took a while to dawn among other physicists, Seaborg grasped them readily. The creation of U-233 from thorium, he predicted, would be "a $50 quadrillion discovery." Once again, though, thorium's potential went untapped: it took Seaborg nearly 20 years to get around to patenting a commercial process for producing U-233. Under patent number 2951023, a body of thorium carbonate is compressed into a dense pellet and placed into the outer blanket of a thermal nuclear reactor. Once the ratio of U-233 to Th-232 reaches about 1:100, the pellet is removed from the core. Voilá: a limitless energy source, bred in a reactor.

TH90 • TH90 • TH90

WITH THAT, LIKE A SWIMMER dropping beneath the waves of a troubled sea, thorium effectively disappeared from mainstream scientific research for six decades or so. Uranium, progenitor of both the Fat Man and Little Boy bombs, took over the show. The discoveries of Curie, Rutherford, Bohr, and Seaborg led to the creation of the largest scientific R&D project ever known, the Manhattan Project. Under the direction of Robert Oppenheimer, scientists raced to build an atomic bomb before Nazi Germany focused exclusively on uranium as the fuel source. Raw uranium was secretly imported from the rich uranium mines at Shinkolobwe in the Congo. At Oak Ridge in Tennessee and Hanford in eastern Washington, great industrial plants were built and run 24 hours a day to achieve vanishingly small amounts of highly enriched uranium (HEU). U-235 and plutonium instantly became the most valuable materials on Earth. By the end of the war, the two remaining superpowers were building cities where they developed bombs to destroy their enemies' cities, and uranium took its place as the only element considered a potential fuel for both bombs and power plants.

There was a brief sideshow, however, that revealed the combatants' burgeoning obsession with radioactive materials and the deep paranoia they engendered. Late in the war, anticipating the fall of the Reich, the United States created an intelligence team, code-named Alsos, to uncover the details of the German atomic bomb program. (Blocked by a lack of resources and indifference on the part of the Fuhrer, that program had fizzled out by 1944. The Allies didn't know that.) Greek for *grove*, Alsos was named for Lieutenant General Leslie Groves, the gruff military head of the Manhattan Project. After the war the Alsos team would cross Germany in search of evidence of nuclear weapons production. Before that, as the Reich held out against the Allied invaders, Alsos agents collected intelligence and rumor in German-occupied Europe.

One of the German companies involved in procuring and processing uranium for weapons research was the Berlin chemicals conglomerate Auergesselschaft, which, among other things, had helped develop thorium mantles for gas lanterns as well as one of the earliest commercially produced lightbulbs, made from the element osmium and wolfram, or tungsten. Auergesselschaft's roots reach deep in the history of nuclear physics; the scientific director, Nikolaus Riehl, had studied under Lise Meitner and Otto Hahn, the discoverers of nuclear fission.

In the fall of 1944, as Allied troops closed in on Berlin, Alsos agents in France learned that the German company had confiscated a French firm, Terres-Rares, soon after the Nazis occupied Paris. Terres-Rares had one of the world's largest stockpiles of thorium. After D-Day the Germans had arranged to ship hundreds of tons of Terres-Rares thorium east into the Reich. When Alsos arrived at Terres-Rares's Paris office after liberation, it was empty. According to Groves, a German chemist named Jansen was in charge of bringing the thorium to Berlin, but he was captured at the French-German border. In his briefcase was a pile of documents that included a file on the city of Hechingen, the center of atomic research in Germany. At Hechingen German researchers under Werner Heisenberg had built an isotope separation unit and an experimental pile—a rudimentary nuclear reactor—in a cave. A ton and a half of uranium was found buried in a nearby field. Adding two plus two to make six, the Alsos field agents concluded that the Nazis had gotten far closer to building an actual bomb than was previously understood. The agents also concluded that the Nazis had found a way to use thorium to make weapons-grade material.[18]

In the postwar years a less alarming explanation emerged: Auergesellschaft officials, realizing that the end of the Reich was near and that selling HEU to the German military was probably not a viable future business model, had decided to diversify, moving into cosmetics and other consumer products. Thorium had already been used to make toothpaste; in fact, Auer had a patent for it and an ad slogan: "Have sparkling, brilliant teeth—radioactive brilliance!"

The case of the purloined thorium had all the elements of a thriller. The real story was more pedestrian, but its coda had important implications for the Cold War. After Berlin fell, the Red Army sent its own detection team after German nuclear secrets. Equipment, technology, and even unlucky personnel were sent east, behind what would come to be known as the Iron Curtain. Along with intelligence from spies inside the Manhattan Project, Auergesellschaft helped the USSR fast-track its own atomic bomb program. The Soviet Union exploded its first nuclear weapon in 1949, a decade earlier than U.S. experts had predicted. The company itself was later taken over by a U.S. corporation, Mine Safety Appliances Corporation.

TH90 • TH90 • TH90

DESPITE ITS PROVEN EFFICACY as a nuclear reactor fuel, thorium was relegated to various industrial uses for the next six decades. Gas lanterns continued to be produced with illuminators known as Welsbach mantles, built from thorium oxide (ThO$_2$) and cerium oxide. Used as a catalyst in a variety of chemical processes, including, ironically, the cracking of petroleum products (the breaking down of complex organic molecules into simpler forms, including light hydrocarbons), thorium is also a key ingredient in the production of high refractive index glass, used in high-end camera lenses. Its extremely high melting point, about 3,300 degrees centigrade, makes it an ideal material for high-temperature crucibles.

All of which is akin to using a Ferrari to drive to the corner market for milk. Despite the wide understanding of thorium's energy potential, despite years of research on thorium and molten salt reactors at one of the preeminent nuclear labs in the United States, despite the evident problems raised by uranium reactors over the decades, a revival of interest in thorium as a source of energy would have to wait until the twenty-first century and the so-called renaissance of nuclear power.

THREE

THE ONLY SAFE REACTOR

A few days after Christmas 2010 the container ship *Altona,* bound for China and carrying a load of 770,000 tons of uranium concentrate—known as yellowcake, the transportable form of uranium that will eventually be processed into fuel for nuclear reactors—was steaming west across the South Pacific. Somewhere between Hawaii and the Midway Islands it ran into a storm. After three days of gales and heaving seas, the crew (like most modern superships, the *Altona* was manned by only a skeleton crew) discovered that the containers in the hold had shifted and two drums of yellowcake had been smashed open. There was loose uranium in the hold.

Owned by Canadian mining giant Cameco, the *Altona* was delivering yellowcake to a Chinese utility in Zhangjiang. (China has become one of the world's largest consumers of uranium.) The ship turned around and headed back to British Columbia, and Cameco executives assured the world press that there was no serious danger of the uranium leaking into the sea or harming the crew. The *Altona* berthed at Vancouver before transferring up the coast to a Cameco plant for clean-up.

The mishap sparked little outcry, joining the dozens of incidents yearly in which uranium in its various forms is mishandled, spilled, mined, or stored in violation of environmental regulations—or simply lost.[1] The *Altona* averted disaster, but it is an apt symbol for the unsettled waters into which the nuclear power industry is sailing.

Until the earthquake-bred tsunami struck the nuclear power plant at Fukushima, Japan on March 11, 2010, nuclear power was enjoying its most promising year in a few decades. The renaissance for conventional nuclear power—based almost entirely on light-water reactors and entirely on the uranium fuel cycle—promised the construction of dozens of new reactors, the addition of several gigawatts of carbon-free power, and a long-delayed windfall for nuclear power producers. China alone has announced plans to add 110 gigawatts of nuclear power-generation capacity before 2021, an amount equal to 25 percent of the total world nuclear power today. Since China's seemingly inexhaustible appetite for coal, the world's dirtiest energy source, threatens to torpedo international efforts to curb global climate change, the shift to nuclear power, which emits no carbon, would be a welcome development—if it could be pulled off.

India plans to add dozens of new reactors by 2050, many eventually powered by thorium (see chapter 7 for a full look at India's thorium power plans). Other countries with ambitious plans for nuclear power plants include South Korea and Japan—although new nuke plants in Japan could well be derailed by the backlash from the accident at Fukushima. Cameco forecasts that more than 100 new reactors could be added worldwide by 2022, further stretching already thin supplies of uranium.

And then there's the United States, where no new nuclear plants have been permitted since the 1970s and none have come online since 1996.[2] U.S. energy producers, particularly the large nuclear utilities Exelon and Duke Energy, have embraced the nuclear renaissance with enthusiasm.

According to the Nuclear Energy Institute, the industry's trade association, 17 companies are planning at least 30 new nuclear plants in the United States. Operators have applied for licenses from the U.S. Nuclear Regulatory Commission (NRC) for 20 plants.

So far, the renaissance has been more heat than power, as it were. Not a single new nuclear plant has won licensing from the NRC. Schemes are afoot to expedite the NRC's famously glacial approval process, which can take a decade or more, but they have not yet borne fruit. Investors looking at the huge upfront capital costs for nuclear power, projected to reach several billion dollars, plus the 10- to 20-year timeline to completion, are hardly rushing to pour money into new nuclear projects. (I'd add, though, that a coal plant also will cost billions of dollars to build—especially as environmental groups and federal agencies succeed in preventing power utilities from passing on the social and environmental costs of dirty coal to ratepayers.)

Several U.S. nuclear power plant projects have been delayed or canceled. In October 2010 Constellation Energy said it was backing out of the Calvert Cliffs nuclear plant project in Maryland. At one time Constellation and its partner, Électricité de France, planned to add a third reactor based on a new design to the two existing reactors at the site, doubling the power output. Constellation quarreled with the U.S. Department of Energy (DOE) about the fees the DOE planned to charge for guaranteeing loans for the $7.5 billion project and finally pulled out.

Six months later NRG Energy killed its South Texas Project, which included plans for two 1,350-megawatt advanced boiling water reactors. NRG cited "multiple uncertainties" regarding funding from investors—the Tokyo Electric Power Company and the Japan Export Bank—stemming, obviously, from the Fukushima incident.

NRG was among the first entrants in the new nuclear derby: the company filed a license application with the NRC in September 2007. Besides investors' second thoughts, the South Texas Project fell victim to another brute fact of new energy programs in the twenty-first century: "It is cheaper, and faster, to build new natural gas fired electrical generation plants than to build new nuclear reactors," wrote the nuclear power journalist Dan Yurman on his blog, *Idaho Samizdat.* "Texas is swimming in a huge surplus of natural gas that sets the price for electricity."

The average two-reactor nuclear plant, based on conventional technology, costs $5–$20 billion to build, once the cost of capital is included. A comparable natural gas plant costs about half as much. You can build a small, 150-megawatt natural gas plant for about $200 million. Meanwhile the world's existing reactors, which include many facilities already years beyond their original planned production life, could be reduced in the coming years. An April 2011 study by the financial services giant UBS found that as many as 30 nuclear plants, including many in areas of seismic activity, could shut down by 2016.

TH90 • TH90 • TH90

THEN THERE IS THE QUESTION OF FUEL SUPPLY. Supplies of uranium, which have been in surplus since the fall of the Soviet Union, have tightened and prices are rising again after plummeting during the world financial crisis of 2008–2009. Anticipation of new nuclear construction has helped drive the price of uranium to levels not seen in years: uranium hit a three-year high of $73 a pound on the spot market at the beginning of February 2011, and there are predictions that it could go much, much higher. A pair of analysts at CRU Group, a commodities and minerals analysis and consulting firm, said in a report in early 2010 that prices in the next decade could challenge the all-time high of $136 per pound, set in the pre-crash days of 2007. As with many energy commodities, the wild card in this calculation is China.

"The current unknown is when and to what extent the Chinese will buy uranium," said Philip Macoun of CRU.[3] The amount of uranium China is buying has surprised most analysts and has helped push the price upward.

Rising prices are good news if you're a producer of uranium, like Cameco or the French nuclear giant Areva, or a uranium-rich country, like Kazakhstan (which in 2010 took over as the world's number one source of mined uranium). If you're a nuclear power utility in a country where developing new uranium supplies has not been a high priority—like, say, the United States—the scenario is less rosy. Fuel costs for existing reactors have been historically low for a long time and comprise a fraction of the cost of nuclear power. One way or another, though, major changes are coming to the uranium market in the next 20 to 30 years. Demand, and prices, are certain to rise; uranium suppliers like Cameco, operating on the theory that "if you look hard enough, you'll find it," are confident that there's enough uranium in the ground to meet any future supply needs. But uncertainty has begun to leak into the market.

Those uncertainties are fueled by two parallel developments: rising demand for freshly mined uranium and the falloff of the so-called secondary market—that is, uranium that has already been produced. After the Iron Curtain fell, the United States began the Megatons to Megawatts program (M2M) to secure and safely reprocess dismantled nuclear warheads in the former Soviet Union. Since the mid-1990s M2M, which was scheduled to end in 2012, has dumped hundreds of tons of uranium onto the world market. Along with existing above-ground supplies, the repurposed Soviet material accounts for nearly one-third of the annual requirements of the world's nuclear plants. That means that the yearly supply of newly mined uranium falls short of demand by 20 to 30 percent. The world consumes on the order of 180 million pounds of uranium a year and produces only about 140 million. With the end of M2M and the depletion of uranium inventories built up in the 1990s and early 2000s, "new primary production will be required," Cameco's CEO, Gerald Grandey, said in the spring of 2011.[4]

Ken Peterson, vice president of nuclear fuels at the Chicago-based Exelon Corporation, the largest producer of nuclear energy in the United States, pointed out to me in an interview that major American utilities are covered by long-term uranium contracts lasting into the next decade. "We have 100 percent [supply] coverage through 2015."

That may be enough for Exelon's executives and investors, but it demonstrates a blinkered view in terms of the longer term and the broader market. The third factor affecting uranium supply is the long lead time for bringing new uranium finds into production: at least ten, and often 15, years. That means there's a doughnut hole in long-term uranium forecasting: although there's plenty of uranium in the ground, the problem is bringing it to market fast enough to meet the needs of power plants, according to Grandey. "Our challenge is investing and permitting and developing new mines fast enough to overcome the drop-off in Megatons to Megawatts," he said at an industry symposium in Denver in the spring of 2011.

Indeed, supply disruptions have occurred recently at big mines, including the Ranger and Olympic Dam deposits in Australia, while several large projects in development, such as Trekkopje in Namibia and Cameco's Cigar Lake mine in northern Saskatchewan, have been delayed.

That's why the rapidly developing nations of India and China, as well as developed Asian countries like South Korea, are moving aggressively to build partnerships and form joint ventures with uranium producers, particularly those in the former Soviet Union. In January 2011 Moscow and Tokyo announced an agreement to jointly prospect for and develop uranium ore deposits, as well as construct and operate new reactors. China, meanwhile, may well be stockpiling uranium: the country reportedly imported about 5,500 tons of yellowcake last year, while it consumes only half that much.

Burned by aggressive supply contracts and failed uranium-mining ventures in the 1970s and 1980s, U.S. utilities—most of which pay more heed to signals from Wall Street than from the world's uranium mines—take a shorter-term, not to say more complacent, view.

The Americans are "out of synch with the rest of the world," says Mariangeles Major-Sosia, the vice president of international network coordination at Areva. "They don't have the longer vision that other countries have, because the U.S. doesn't really have an energy policy. They're publicly traded companies, so they're making decisions to next year or next quarter," she told me when we met at the symposium in Denver.

As long as the current generation of utility executives remains in power, that's unlikely to change. The renaissance waits upon a change not only in energy policy and economics but in leadership.

TH90 • TH90 • TH90

CHARLES HESS WAS NEVER one of those executives, in the corner suite and chauffeured town car sense of the word, but he has spent 35 years in the nuclear power business as a chief engineer, research director, and plant operator. He lived out the contraction era, and he is now living through the prospective new age of expansion. And he has come to believe that thorium is nuclear power's inevitable future.

In March 1979 Hess was working at a nuclear plant in New Jersey. The Three Mile Island accident started before dawn on March 28. That morning he and his fellow engineers at Oyster Creek Generating Station in Ocean County heard about the accident on the radio, and later that day he was devising ways to deal with the radioactive gas that was accumulating in TMI's main reactor building and the neighboring auxiliary building. He spent the next six months as one of the lead engineers trying to contain the radioactive spillovers from the accident. The experience profoundly redirected Hess's career as well as the nuclear power industry in the United States.

Specifically, Hess worked on a method to transfer the highly radioactive standing water in the reactor building to tanks in the spent fuel pools nearby. This was not a simple operation; in fact, nothing like it had been done before. The job was complicated by internal politics: essentially Hess and his colleagues were treated as if they were responsible for the accident. Hess developed eductors (also called jet pumps or Venturi pumps) to remove steam from the storage tanks. Ultimately, none of the systems that Hess designed and built at Three Mile Island was ever used. This would become a common theme in his career and that of many other long-time nuclear technologists': work for years on a project, or a reactor technology, only to see it delayed, mothballed, and finally canceled. The nuclear power industry has been full of dead ends since the early 1980s.

Hess spent much of the 1990s working on the next generation of advanced light-water reactors (new versions of the uranium-based reactors that already dominated the industry) for the Department of Energy. Designed to be the third generation of nuclear reactors, the advanced designs included a breeder reactor cooled with liquid metal from GE. Liquid metal breeder reactors were the ultimate dream of the nuclearati; these machines would run on spent fuel from conventional light-water reactors while creating more fuel and burning up the long-lived isotopes that in nuclear waste remain radioactive for thousands of years. These advanced breeder reactors made previous plants look as outmoded as nineteenth-century coal furnaces. On paper, on computer screens, they were elegant, possessing the beauty of highly engineered machines. They were the products of the finest nuclear technologists in the world, and R&D on them totaled billions of dollars in the 1980s and 1990s. While a few have been built and have supplied power to the grid, they are inconsequential as a percentage of total nuclear power. The liquid metal breeder program was finally canceled by the Clinton administration, which killed all nuclear and fossil power programs at the DOE. By the mid-90s, government-supported research and development on new nuclear technology in the United States had effectively halted.

TH90 • TH90 • TH90

SLOWLY, IN THE LATE 1990S, the nuclear power industry began to show signs of life. In 1998 Entergy Corporation, the second-largest producer of nuclear power in the country behind Exelon, bought the Pilgrim nuclear station on the coast of Massachusetts. It was the first time in more than a decade that an American company had willingly expanded its nuclear assets, and it set off a wave of other deals that culminated in a bidding war for the Millstone plant, the only nuclear power station in Connecticut. Sold in 2000 to Dominion Resources, Millstone was purchased for about $800 per kilowatt of capacity, about half of the cost of a new nuclear plant. That sale convinced a lot of insiders, like Charlie Hess, that nuclear was back. By the turn of the century, the industry had found a new business model that didn't require expensive new builds: extend the life of existing plants, "uprate" them to run at close to 100 percent capacity, refinance debt, take advantage of rising prices for electricity, and sit back and watch the cash roll in. When George W. Bush took office in January 2001, no one in this country and hardly anyone in other countries was seriously considering building a new nuclear power plant. When Bush left office eight years later, the NRC had more than a dozen applications for construction and operating licenses for more than 30 plants.

Hess took a job with the International Nuclear Safety Program, the progenitor of M2M. He helped Russian nuclear plants, which are not exactly known for their sophisticated safety mechanisms, improve their shielding and containment systems. It was work he liked, collaborating with men who had, earlier in their careers, studied ways to obliterate U.S. cities. When he came back to the States, he joined the Shaw Group, a big engineering and construction firm that includes a high-end nuclear consulting arm. Hess became the project manager of Shaw's Next-Generation Nuclear Plant program, helping plot the technological road map for the twenty-first century. The program's main focus is the pebble bed reactor, another design that has been on the books for decades. Pebble bed reactors offer safety and efficiency improvements over conventional light-water designs. But by now Charlie Hess had seen the future of nuclear power, and it was based on thorium.

"It is clearly the next step," he told me in one of our phone conversations. "I think first we'll see sodium-cooled fast reactors, and after that thorium is the obvious choice."

With the election of Barack Obama, the window of possibility for alternative nuclear energy opened a bit wider. Obama favored nuclear power but for political reasons was adamantly opposed to storing nuclear waste at the Yucca Mountain facility in Nevada. Soon after his inauguration he formed the so-called Blue Ribbon Panel to study the problem of long-term waste storage. The new energy secretary, Steven Chu, had been vocal in his support for nuclear power, particularly advanced reactor designs. Charlie Hess was asked to head up a DOE subcommittee charged with making recommendations on technology roadmaps going forward. The nuclear renaissance seemed to finally be reaching critical mass (a popular metaphor among the nuclearati).

In early 2011 Hess's team recommended two possible pathways: the pebble bed reactor and a thorium-fueled fast reactor that would convert thorium to plutonium fuel. It was, to Hess's knowledge, the first time since the 1960s that an official government-sanctioned body had recommended pursuing thorium power.

Then came Fukushima, and the future of nuclear power was once again cast into doubt. Although the industry was approaching record profits, anyone with a creative approach or an alternative design had long since chosen to leave, or been driven out of, nuclear power. Why take a risk on a new fuel, even one with proved advantages over uranium—in safety, environmental concerns, and unsuitability for conversion to weapons—when the industry was finally in the black?

Nevertheless, the thorium revolution "is certainly going to happen," said Hess. "It's happening right now in India, and it's going to happen in a lot of other countries."

The United States has already ceded its leadership position in nuclear R&D. The future of innovation belongs to international consortia of megacorporations and perhaps to agile, risk-taking entrepreneurial start-ups. "It's a new world," said Hess.

TH90 · TH90 · TH90

THE NEW WORLD WILL REQUIRE NEW APPROACHES, not only to reactor design but to the nuclear fuel cycle and the entire art and science of nuclear power. The "if it ain't broke, don't fix it" complacency represented by long-time executives like John Rowe of Exelon was definitively killed off by the Fukushima-Daiichi accident. To be sure, the nuclearati believe they have the technology to support the nuclear renaissance. At least a dozen so-called Gen III and Gen IV reactor designs are currently being studied, and in some cases they actually have been funded, but they all share a common flaw. They're all based on uranium.

Here I'll review the basic structure and operation of a nuclear power reactor. In its simplest form a reactor is nothing more than a machine for generating heat, which in turn is used, most often by way of a steam-powered turbine, to make electricity. In the core a nuclear reaction—sustained in U-235, in plutonium, in thorium-born U-233—generates energy in the form of heat. The core heats water, which generates steam. The steam turns a turbine, converting the kinetic energy of the superheated fluid to mechanical energy. The turbine spins a set of windings inside an electrical generator that converts the mechanical energy into electricity. In its outline this description could fit any form of turbine, whether turned by falling water, burning coal, or natural gas. Heat source, steam, turbine, generator. Electricity from simplicity.

The dangerous and fickle nature of radioactivity, though, adds several complexities to this simple system; three are the most significant. The first is that, unlike, say, a coal-fired power plant, you can't just turn off the system by depriving it of fuel. Apart from the residual heat from coal already in the burner, a coal plant without fuel shuts off and begins to cool immediately. Once you start a nuclear reaction, you can't just shut off the heat. You can quickly reduce the fission reactions by inserting control rods that slow down the reaction, but the production of residual heat from the decay of fission products keeps going until all the material in the core has decayed to stable isotopes—that is, for a very long time at a very high temperature. That's why the Pandora's box and genie-in-a-bottle metaphors have appealed to antinuclear activists since the dawn of the Atomic Age. The nuclear power bottle is hard to recork.

The second complexity is in the plumbing. Light-water reactors come in two main versions: pressurized water reactors (PWRs) and less-common boiling water reactors (BWRs). Both operate under high pressure: 2,200 to 2,500 pounds per square inch (psi) in PWRs and 1,000 to 1,100 psi for BWRs. Circulating water and steam from reactor to turbine under such high pressures requires a complicated network of pipes, valves, and other plumbing that can fail, corrode, or fall prey to operator error. What's more, in PWRs the radioactive primary coolant does not actually generate the steam that drives the turbine. A secondary cooling system, with a second heat exchanger called a steam generator, produces the steam that turns the turbine. Naturally this extra layer reduces the efficiency of the system. More important for this discussion, it adds to the complexity of the machine, nearly doubling the sheer volume of plumbing and increasing the opportunities for failure and explosions.

Secondary cooling systems are not unique to solid fuel reactors; most liquid fuel designs require some form of secondary liquid. But early in the research into nuclear reactors it became apparent to scientists that a liquid-core reactor would greatly simplify the overall system. Among the leaders of the Manhattan Project, this concept appealed most forcefully to Eugene Wigner. Wigner's ideas gave birth to the Molten Salt Reactor Experiment, described in detail in chapter 6.

To return to conventional reactors, the third complexity is the moderator. Uranium-fueled reactors require some type of moderator to slow down the scattering neutrons enough to sustain the fission reaction. Many, many materials have been used to moderate commercial reactors: ordinary water, heavy water, solid graphite, and so on. The choice of moderator determines many other characteristics of the reactor: type of fuel, type of fuel cladding, operating temperature, the metal used in the core vessel, and the like. No moderator is ideal; all have their advantages and disadvantages. Uranium is like a finicky child at a buffet: only the right combination of moderator, fuel, core design, and materials will produce a sustained fission reaction, and even then the ratio of inherent energy in the fuel to electricity produced is low.

In general, it must be said, the light-water reactor has worked as advertised. Despite a few notorious accidents (Three Mile Island, Fukushima-Daiichi) and one genuine disaster (Chernobyl), the overall safety record of nuclear power is quite good—which nuclear power executives, as tone-deaf to public mistrust as any group of business leaders outside the tobacco industry, never tire of pointing out. Once built, nuclear power plants produce power as inexpensively as any energy source (a statistic underlined by the low price of uranium during the industry's Dark Age, from 1980 to the early 2000s). Uranium mining, the effects of which ravaged many Native American communities of the American Southwest in the middle decades of the twentieth century, is still opposed in many places on environmental grounds. But by the limited standards of the nuclearati, nuclear power is a success.

From a wider perspective nuclear power has failed dismally to fulfill its promise. In his famous "Atoms for Peace" speech to the United Nations in 1954, President Eisenhower touched on the promise of the Atomic

Age: "Who can doubt that, if the entire body of the world's scientists and engineers had adequate amounts of fissionable material with which to test and develop their ideas, this capability would rapidly be transformed into universal, efficient, and economic usage?"

Universal it is not. Nuclear power today produces only about one-fifth of the electricity in the United States; in a few countries, like France, that figure has reached four-fifths, but they are exceptions, and several countries, including Germany, which has Europe's largest economy, have said in the wake of Fukushima that they will abandon nuclear power altogether. In Japan nuclear power provides 30 percent of the nation's electricity; how that will change in the post-Fukushima era remains to be seen.

Nor is nuclear power economical. Nuclear plants are now so expensive to license, not to mention build, that there is little prospect of investors' receiving reasonable returns. Many, many words have been written about the "true costs of nuclear power." Much of this analysis is based on faulty reasoning or is outright bogus, but there is no question that the rosy pronouncements of industry groups like the Nuclear Energy Institute are specious, given the true social costs of producing power from uranium.

Nuclear power has also failed to demonstrate that it has the third of Eisenhower's predicted attributes. No power source that leaves behind 95 percent of its fuel's available energy, in a form that remains intensely toxic for tens of thousands of years, can be considered efficient. The problem of nuclear waste in the United States is insoluble under current political and market conditions. (In other countries this is not so: France has been recycling its spent fuel for decades). The position of the Obama administration—supporting the development of advanced nuclear power while effectively killing off the Yucca Mountain disposal facility—is self-canceling. Without some long-term, politically acceptable method of storing nuclear waste (or, better yet, vastly reducing the volume and radiotoxicity of the waste in the first place), none of the optimistic forecasts of the nuclear renaissance will come true.

The longer I reported on nuclear issues, the more I found myself holding two sharply conflicting opinions: on the one hand the conclusion that only nuclear power can rescue us from the cycle of dependency on dwindling fossil fuels and destructive climate change in which we're trapped, and on the other a deep frustration with the nuclear power industry as it is constituted.

The biggest barrier to the revival of thorium power has nothing to do with thorium itself. The biggest barrier is the intransigence and complacency of the industry itself and the inability of our political leaders to reform it.

TH90 • TH90 • TH90

URANIUM'S SHORTCOMINGS AS A SOURCE of energy have been evident since the days of Madame Curie, plain to everyone who has worked with it, but its advantages, both as a reactor fuel and a bomb-making material, have hidden the shortcomings in much the same way that uranium itself has overshadowed thorium. The shortcomings fall into four main categories.

The first flaw is uranium's natural state: as Bohr realized on the eve of World War II, only the smallest fragment of uranium ore, the 0.7 percent U-235, is actually fissile. The rest is radioactive but relatively stable U-238. You can produce energy from unenriched uranium—Canadian companies have been doing it for decades—but it requires more sophisticated reactors than the United States has, and the electricity is more expensive. Enrichment is an expensive process that requires industrial scale and an advanced level of technology.

The second flaw is that, once uranium has been refined enough to use it in a nuclear reactor, it can be used to make nuclear weapons. No one has ever successfully constructed a bomb using smuggled or stolen uranium, but that is not for lack of trying. Entire ministries and government agencies have as their sole raison d'être the control of the traffic in enriched uranium; though the danger of losing, misplacing, or being robbed of the stuff is admittedly small, it adds greatly to the expense, risk, and the public opprobrium of running nuclear reactors.

Uranium's third flaw is that, even though it's such an efficient source of energy compared with coal and other carbon-based fuels, current nuclear plants use only 3 to 5 percent of the available energy in a given amount of low-enriched uranium. The rest is thrown out as spent fuel, presenting a disposal problem that the industry has never fully solved.

The final flaw is more complicated, and it has to do with the basic nature of uranium atoms and the inherent processes of nuclear fission. In brief, in nuclear engineering language, you can't build a breeder reactor in the thermal spectrum using uranium; a uranium breeder must be a fast reactor.

To uncoil that sentence, recall the batter analogy from chapter 2. The slower the pitcher's average velocity, the greater chance the hitter will connect with any given pitch; as average velocity rises, the chances of a hit go down, rapidly. In a conventional light-water reactor, a moderator—usually plain water, sometimes graphite—is used to reduce the energy of neutrons careening off from nuclear fissions. It's as if between the pitcher's mound and home plate there were an invisible barrier that slows the baseball without stopping it altogether.

The equivalent of batting average in a nuclear reactor is known as the cross section: the probability that any neutron will interact with another atom and cause it to fission. (There's also a cross section for neutron *capture*, which is not what you want in a reactor. Capture and fission are inversely related; as more neutrons are captured, fewer atoms split, reducing the chances of prolonging the fission reaction.)

In thermal reactors, which account for the vast majority of nuclear plants worldwide, neutrons travel at speeds of about 2,200 meters per second. In fast reactors, they move much, much faster, reaching speeds of 9 million meters per second. That's about 336,000 miles per minute, or about 3 percent of the speed of light. In other words it's about 17 million times as much energy, the increase gained simply by removing or changing the moderating material. Water, which contains two hydrogen atoms per molecule, is an excellent moderator because hydrogen atoms are good at slowing down neutrons. Fast reactors, on the other hand, typically use heavier elements, like sodium, which is much less of a speed bump for neutrons. (Paradoxically, lighter atoms are more effective at slowing down atoms. To understand why, picture a pool table with a bowling ball on it. If the cue ball strikes the eight-ball, the cue ball will transfer most of its energy to the other ball, often stopping dead. If the cue ball strikes the bowling ball, on the other hand, it will carom off at almost the same speed at which it was traveling before the collision.)

Uranium-238, the most common form of natural uranium, will actually fission at sufficiently high energies (typically more than two million electron-volts). In physicists' parlance, it is fissionable but not fissile; it will split apart only when struck by a neutron traveling very, very fast.

In a fast reactor, the fission cross section goes down, but the breeding ratio goes up (in technical terms the "neutron flux"—the density of neutrons zipping around—is higher). The breeding ratio is simply the ratio of the fuel produced to the fuel consumed. A liquid metal fast breeder, for example, is fueled with uranium and plutonium. Its breeding ratio is between 1.1 and 1.4 to 1, meaning it produces 10 to 40 percent more fuel than it burns. That's possible because the number of neutrons released per fission is high: well above two. Because of the high number of neutrons released in the fast fission of plutonium-239, it's possible to not only sustain the fission reaction (the burning of U-238) but also to create more PU-239 than you are consuming. The reactor makes up in volume what it loses in efficiency.

Put another way, a fast breeder is like a party for 500 college students with plenty of blaring music, strobe lights, and alcohol. Put that many young people in the same place with that much sexual energy, and you're bound to produce a certain number of hook-ups. At the same time, the potential for mishaps is high.

One of the earliest fast breeders, the Sodium Reactor Experiment at Santa Susana Laboratory, in California, suffered a partial meltdown in July 1959, when a so-called power excursion caused the core to overheat. A third of the fuel melted, and a significant amount of radioactive gas was vented into the atmosphere. Seven years later the Fermi–1 breeder in Monroe, Michigan, partially melted down after a fragment of zirconium cladding, which lined the meltdown cone, got stuck in a pipe in the cooling system. Fast breeders are "twitchy beasts," as Kirk Sorensen likes to say, for a few reasons: primarily because events inside the reactor are simply sped up, and dramatic changes can occur instantaneously, reducing the margin for error.

Second, the most common coolant in fast breeders, liquid sodium, explodes when it comes into contact with air or water. When Fermi–1 was restarted in May 1970, a sodium explosion damaged the core, delaying things another few months. The twitchy plant shut down two years later after its operating license expired and was not renewed. The exotic materials needed to safely run fast breeders require a level of diligence and sophistication that the nuclear power industry has not always exhibited. That hasn't stopped the industry, and the government-funded national labs, from pursuing fast breeders at a cost of thousands of human work hours and billions of dollars. Thorium reactors lost out to light-water reactors as the industry's original choice for its workhorse reactor; then thorium reactors lost out to fast breeders as the choice for the future of advanced reactors, despite the questions and objections that dogged fast breeders throughout the 1970s and 1980s.

"The need for the holy grail that drove this whole business, namely the fast breeder reactor, became subject to searching questions that could not be satisfactorily answered, . . . and not only on grounds of safety and proliferation."[5] Not only were fast breeders dangerous and difficult to operate, but they would take a long time to justify their high costs: as much as a century to reach "doubling time," the amount of time required to produce enough excess fuel to start another reactor of the same size, even if you could keep fast breeders running without everything going kerflooey. Today's designs for Gen IV fast breeders claim a doubling time of 20 years or less, but that's on paper.

Expense and inherent safety have never been primary requirements for the next revolution in reactor technology; the industry's pursuit of fast breeders has gone on for more than 60 years and continues today. In 2010 the Monju reactor in Japan, a fast breeder that was shut down for almost 15 years after a 1995 fire caused by a leak in the secondary coolant, was shut down again after a refueling accident. Japanese officials, who mounted a strenuous cover-up in the years immediately following the accident, said the reactor would not be restarted until at least 2014. But they were not giving up.

A 2010 report from the International Panel on Fissile Materials noted that more than $50 billion had been spent worldwide on fast breeder R&D, including more than $10 billion each by Japan, Russia, and the United States. "None of these efforts has produced a reactor that is anywhere near economically competitive with light-water reactors," the report concluded. "After six decades and the expenditure of the equivalent of tens of billions of dollars, the promise of breeder reactors remains largely unfulfilled and efforts to commercialize them have been steadily cut back in most countries."

So much for the holy grail.

TH90 • TH90 • TH90

SO, IF YOU WERE GOING TO DESIGN and build a new nuclear reactor from scratch, what would it look like?

First of all, you'd make it small. The old antinuke saw says, "Nuclear reactors come in only one size: extra large." But compact modular reactors that can be prefabricated, transported by shipping container, and assembled on site are now seen by many experts as the future of nuclear energy.

"If you go small, and manufacture reactors like Henry Ford did cars, there's a host of advantages," Tom Sanders told me shortly before he took over as president of the American Nuclear Society in 2009. (He is now its president emeritus.) "You could use automated manufacturing processes instead of doing every weld individually, you could get the plants licensed in a two-year time frame instead of seven, and it'd be much cheaper on a per-kilowatt basis."

Virtually all the major nuclear vendors, including GE-Hitachi Nuclear Energy, Bechtel (a company not exactly renowned for miniaturization), Babcock & Wilcox, and Westinghouse (now owned by the Korean tech giant Toshiba) are developing small modular reactors (SMRs). These reactors can use uranium or thorium (or even plutonium), but thorium, with its higher efficiency, offers unique qualities that make it well-suited for miniaturization. They produce less than 300 megawatts, the limit for an officially small reactor. Future versions that could fit on the back of a flatbed truck are envisioned at 60 or even 30 megawatts. Like mobile homes, SMRs can be manufactured centrally and assembled on site, facilitating financing and shortening the time to production; in theory, multiple SMRs could be combined to create a large generating station. Keeping the plants small and dispersed, though, makes them less tempting targets for would-be terrorists—as does fueling them with thorium.

More important, they could produce energy at a lower price per kilowatt than conventional nuclear plants, bringing the cost of nuclear power more into line with low-cost coal production. Newly infatuated with what's known as distributed power generation (lots of smaller reactors scattered in lots of places), the nuclear industry has finally realized that bigger is not always better.

More compact and more affordable are good things; even better is the prospect that thorium-powered SMRs could help solve the problem of nuclear waste storage and disposal. Some ambitious nuclear designers have even started to dream up small, modular fast breeder reactors, which is a bit like trying to control a tiger by putting it in a smaller cage.

Bringing these designs into commercial production could take a decade or more. The three main barriers to widespread deployment, as Philip Moor puts it, are the same that face any new nuclear plant: "Dirt, licensing, and money," he told me. Moor heads up a special committee of the American Nuclear Society formed to examine the business and manufacturing issues around SMRs.

The Savannah River Site, a nuclear industrial complex operated by the DOE near Augusta, Georgia, will supply the dirt (the real estate and infrastructure), and industry heavyweights like GE, Westinghouse, and Bechtel are lining up to provide the money, at least for demonstration projects. That leaves licensing.

"Once we start the demonstration projects, we can start pursuing the license application," said Sanders of the American Nuclear Society. But "we need something operating on the ground."

That's hardly a slam dunk. It's worth noting that building minireactors is not a new concept. GE actually started the Power Reactor Innovation Small Modular (PRISM) program back in 1981, and in 1994 the NRC issued a report that said the commissioners foresaw no impediments to licensing. The project was abandoned in 2001 and then got a second life in 2006. With huge new supplies of natural gas starting to reach the market, and coal plants still the least expensive form of power generation, new nuclear plants will continue to look expensive. And investors looking back at 30 years of nuclear dead ends are sure to be wary of new technological marvels, however promising. The history of nuclear power demonstrates that nothing is truly viable until the core starts chain-reacting.

Still, thorium-powered SMRs offer the best way forward for new nuclear power—and a potential solution for global warming. Smaller is beautiful, and in this case it could be more profitable as well.

TH90 • TH90 • TH90

SECOND, YOU'D MAKE YOUR NEW REACTOR a breeder, preferably a thermal breeder. The failure of fast breeders to fulfill their promise has not erased their appeal; it has just caused the quest for a fast breeder to go in (slightly) new directions. Breeders would be advantageous not only because, theoretically, you'd never run out of fuel, but also because you can use them to process nuclear waste from conventional reactors. At least in the United States, the question of how to store nuclear waste has no clear answer, and there may not be one for the next decade. Building self-sustaining breeder reactors would, as the nuclearati like to say, "close the fuel cycle"; little radioactive material would be left over to dispose of.

Then you'd want to make your reactor inherently safe. Inherent safety—not to be confused with passive safety, a very different thing—is a term much beloved by nuclear engineers.[6] It has been applied to just about every reactor design, including the uranium-fueled light-water reactor and the sodium-cooled fast breeder, machines whose inherent safety is, to say the least, questionable. Traditionally, the solution to this problem has been external safeguards, also called overengineering: add more controls, more redundancy, more miles of piping, more plumbing and alarms and sensors and gauges, and the inherent twitchiness of the world's most volatile energy source could be contained and controlled. Unfortunately, all that engineering brings more complexity, and complexity in itself adds risk. Virtually all the reactor accidents that have ever occurred have had one of two causes: either a fiendishly complex mechanism failed because of a simple mishap (like a loose chunk of zirconium) or a human being failed at the task of monitoring and managing a fiendishly complex mechanism.

The only truly inherently safe reactor is a liquid-core reactor, like the molten salt reactor that was created at Oak Ridge in the 1960s. For the purposes of a reactor designer, liquid—whether it's water, liquid metal, or some type of liquid fluoride—has a marvelous characteristic: it expands rapidly when it gets hot. All materials expand when heated, of course. In a liquid-core reactor, as the energy of the liquid rises, it expands and naturally, passively, slows down the reaction, making a runaway accident nearly impossible. In technical terms, this is known as a "negative temperature coefficient of reactivity." That means that as the temperature rises (which typically is what happens when something goes wrong in a nuclear reactor), the reactivity goes down. When the reactivity goes down, the reactor is essentially turning itself off.

Liquid fuels have several other characteristics that make them safer than conventional solid fuel reactors. This is where the benefits of thorium, which for a variety of reasons is uniquely well suited to liquid fuel reactors, extend beyond the nature of the element itself. No matter how you use it—in a light-water reactor, in a pebble bed reactor—thorium offers advantages over uranium. But in a liquid fuel reactor, that advantage is magnified. If you put high-octane gas in a 1975 Ford Pinto, you'll see some marginal performance enhancement. To get the full benefit, though, you should put it in a Ferrari Testarossa. Using thorium in a liquid fuel reactor is similar: its unique qualities as an energy source are fully exploited.

For example, in liquids—particularly in molten salts—fission products tend to be stable, making it easier to isolate and remove them. One of these fission products, xenon-135, is a nuclear poison that tends to build up in conventional reactors, slowing down the reactions. It renders the fuel unusable after only a small percentage of the potential energy has been used, and it's hideously difficult to handle as part of the nuclear waste stream. In fluid fuels, because xenon forms a noble gas (one that is impervious to chemical reactions), xenon is easy to remove. In a LFTR it can be boiled off as a gas and processed while the reactor continues operating, reducing downtime and increasing the amount of the potential energy that can be extracted from the thorium fuel. A ton of thorium can produce energy equivalent to that produced by 200 tons of uranium in a conventional light-water reactor.

Liquid fuels are also impervious to radiation damage, solving one of the thorniest problems in solid fuel reactors. Continuous bombardment by neutrons over periods of weeks or months wears down not only the solid uranium pellets in a light-water reactor but also the cladding (usually made of zirconium) that contains them. Because of radiation damage and the buildup of fission poisons like xenon, fuel rods age quickly; they have to be replaced every few years, even though only 3 to 5 percent of their energy has been consumed.

Liquid fuels have one other characteristic that makes them ideal for reactor cores: they flow. Gravity, not elaborate control systems or so-called passive safety systems, gives LFTRs their ultimate protection against a serious nuclear accident. In a criticality accident (i.e., if the fission reaction in the core starts to get out of control), a specially designed freeze plug in the reactor vessel melts and the liquid core simply drains out of the reactor into an underground shielded container, like a bathtub when the drain plug is pulled. The fission reactions quickly cease, and (thanks to the expansive quality noted earlier) the fluid cools rapidly. Decay heat is contained harmlessly. Meltdown is impossible, and there are no solid fuel rods too radioactive to remove.

Inherently safe, LFTRs pose less threat than light-water reactors, coal-fired power plants, oil refineries, or just about any other form of large energy or chemical plant. Built small and modular, they will be less expensive to build and operate than just about any other energy source.

TH90 • TH90 • TH90

FINALLY YOU'D FUEL YOUR SMALL, breeding, inherently safe, liquid-core reactor with thorium. I mentioned in chapters 1 and 2 many of thorium's sterling qualities as a nuclear fuel; they bear reviewing.

It is abundant. In fact, used properly, it's effectively inexhaustible.

It requires no special refining or processing beyond purifying it from the monazite ore in which it is most commonly found. It can be mined safely, with none of the tailings and other results of uranium mining that, in the early years of the Atomic Age, poisoned whole communities in Russia and the United States.

It's no good for making weapons. In fact, it's not fissile at all. It requires a kind of nuclear alchemy to be transmuted into uranium-233, which is a more efficient and safe source of energy than U-235.

Finally reactors based on thorium—or, rather, U-233, into which thorium transforms in a nuclear reactor—consume far more of the latent energy trapped inside the fuel, vastly reducing or even eliminating the problem of nuclear waste.

In short, you'd build a liquid fluoride thorium reactor, or LFTR. LFTRs are the first truly revolutionary reactor design to come along since the development in the 1960s of the molten salt reactor, progenitor of the LFTR. LFTRs are designed with an outer blanket of liquid fluoride that contains dissolved thorium-232—thorium tetrafluoride, to be precise (a fluoride is simply a combination of fluorine and another element; *tetrafluoride* means four atoms of fluorine). The thorium is borne in a solution of lithium and beryllium fluorides that has maximum heat-transfer properties, making it a supremely efficient coolant. This radioactive cocktail surrounds a core of uranium-233 that is produced from the natural decay of Th-232 bombarded by neutrons.

The neutron source, to start the reaction, is typically a small amount of fissile uranium, although the neutrons can also come from a particle

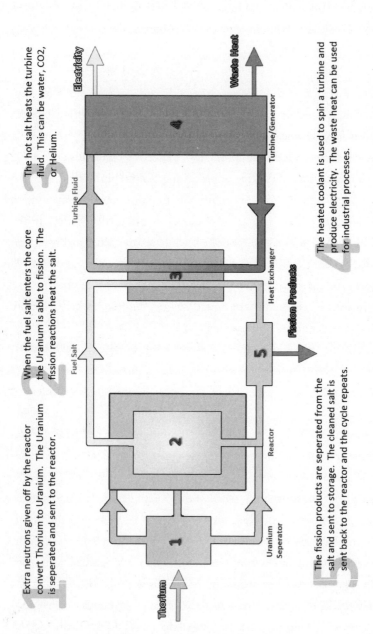

1 Extra neutrons given off by the reactor convert Thorium to Uranium. The Uranium is seperated and sent to the reactor.

2 When the fuel salt enters the core the Uranium is able to fission. The fission reactions heat the salt.

3 The hot salt heats the turbine fluid. This can be water, CO_2, or Helium.

4 The heated coolant is used to spin a turbine and produce electricity. The waste heat can be used for industrial processes.

5 The fission products are seperated from the salt and sent to storage. The cleaned salt is sent back to the reactor and the cycle repeats.

Thorium

Electricity

Waste Heat

Turbine Fluid

Fuel Salt

Turbine/Generator

Heat Exchanger

Fission Products

Uranium Seperator

Reactor

In this simplified diagram of a liquid fluoride thorium reactor, thorium is converted to uranium-233, which sustains the fission reaction, heating a secondary liquid that powers a turbine to create electricity. (Brad Nielsen)

accelerator, of the sort used in physics experiments to smash particles together. The blanket and inner core are in two concentric containers. It's essentially a double boiler: the inner core, sheathed in an exotic alloy of a metal
such as zirconium, contains the fissile U-233, and the outer shell, or blanket,
contains the fertile thorium.

Once the reactor core goes critical, the fission reactions in the core continuously throw off neutrons that keep the thorium, in the blanket, in a
constant state of transformation, creating a virtuous cycle. Such a plant has
two separate loops of piping: one carries the fertile thorium tetrafluoride salt,
once it has been sufficiently bombarded to start the decay chain, into a decay tank from which U-233 can be transferred to the inner core; the other
sends the hot U-233 salt from the core to a heat exchanger to drive a steam
turbine.[7] There are several variations on this basic design, which use various
fluids to transfer heat from the reactor core to the turbine; suffice it to say that
whichever is chosen, it will be significantly more efficient than a conventional
nuclear plant.

After passing through the heat exchanger, the second loop, carrying hot
U-233 fuel salt, cycles back into the core, with a small secondary side stream
passing through a reprocessor, where the fission products are removed, preventing them from poisoning the reaction, before being cycled back into the
core for further fission reactions.

Because the core is liquid, it operates at atmospheric pressure, meaning
that the extremely thick-walled, pressurized vessels used in conventional reactors, which have an unfortunate tendency to blow their top, are unnecessary.
Because LFTRs consume virtually all their nuclear fuel, the majority of the
waste products are not long-lived fissile material but rather fission products,
about 83 percent of which are safe within a decade. While LFTRs, like every
other nuclear reactor, generate fission products that are highly radioactive,
their half-lives tend to be measured in dozens of years, not thousands. The
long-lived radioactivity of LFTR waste is one ten-thousandth that of a conventional reactor. The leftovers, a small fraction of the waste produced by
conventional reactors, must be stored in radiation-proof geological sites for
about three centuries, compared with ten thousand years for nuclear waste
from conventional uranium reactors. In fact, LFTRs themselves make great
garbage dumps for spent nuclear fuel: they can refine standard nuclear waste

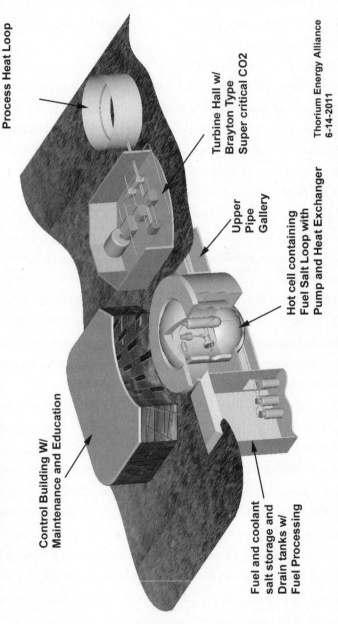

Thorium Energy Alliance
Molten Salt Research Reactor

Process Heat Loop

Turbine Hall w/
Brayton Type
Super critical CO2

Upper
Pipe
Gallery

Hot cell containing
Fuel Salt Loop with
Pump and Heat Exchanger

Thorium Energy Alliance
6-14-2011

Control Building W/
Maintenance and Education

Fuel and coolant
salt storage and
Drain tanks w/
Fuel Processing

This schematic shows a full thorium power plant including a reactor vessel, drain tanks, and a Brayton-cycle turbine using supercritical carbon dioxide. (Thorium Energy Alliance)

into LFTR by-products, essentially solving the currently intractable toxic waste storage problems that plague today's nuclear power industry.

With their high negative temperature coefficient, LFTRs are impervious to sudden overheating. They're also exquisitely tunable; the concentration of fuel in the outer blanket can be adjusted continually, making it easy to control the reactivity in the core.

Finally, they can run practically forever. The reactions in a LFTR produce enough excess neutrons to breed their own fuel. LFTRs are the only type of reactor that can breed more fuel than they consume in the thermal, or lower-energy, spectrum. They have the virtues of fast breeders without the volatility.

Here it is useful to think back to the nature of fission and neutron absorption. In today's conventional reactors, the great majority of the fuel is U-238, which transmutes to the transuranic element plutonium-239 when the U-238 absorbs a single neutron. Thorium-232, by contrast, requires five neutrons to become a transuranic (neptunium-237, which can be safely burned down, or processed, in the reactor). That too makes LFTRs inherently safer than solid-fuel uranium reactors.

While liquid-core reactors can be built to operate without moderators, in some LFTR designs the core does use moderators—typically graphite rods, just as in a conventional uranium reactor. Just as the LFTR has unique qualities that make it superior to light-water reactors, though, U-233 has some distinct advantages over uranium-235, the fissile material that runs the vast majority of the world's nuclear power stations today. U-233 displays a quality that nuclear engineers love: high neutron economy, usually expressed as η in physics equations. That means that an atom of U-233, after absorbing a stray neutron and fissioning, produces on average 2.16 neutrons. Since one neutron is required to continue the chain reaction, 1.16 neutrons are freed up to produce new fuel.

Overall, LFTRs are 200 to 300 times more fuel efficient than standard reactors. They are safer, simpler, smaller, less expensive to build, and less expensive to run to produce electricity on a cost-per-kilowatt basis. No rational from-scratch approach to nuclear power would build anything else, yet we are burdened (some would argue stuck) with a supply of costly, overengineered, unsafe uranium reactors that produce tons of long-lived nuclear waste. All of which raises these questions: How did this happen? Why were thorium-based molten salt reactors abandoned when they showed such promise at the dawn of the Atomic Age?

The answers, as they say, are complicated. And the history of the nuclear power industry is entwined with the destinies of two brilliant and very different men: Alvin Weinberg and Hyman Rickover.

FOUR

RICKOVER AND WEINBERG

O n August 9, 1945, the American bomber *Bock's Car* flew over the southern island of Kyushu and dropped an atomic bomb that engulfed the city of Nagasaki in a nuclear fireball. Fat Man, as the bomb was nicknamed, was 60 percent more powerful than the one that had devastated Hiroshima three days earlier. At that time Hyman Rickover was an obscure Navy captain in charge of the ship repair base at Baten Ko on Okinawa, where *Bock's Car* stopped briefly for refueling on its flight back to the American base at Tinian. Hardly a World War II hero, he would become a major actor in the Cold War, the father of the nuclear submarine, the Navy's longest-serving admiral, and an unrelenting hawk whose belief in absolute, overwhelming nuclear force—based on the nuclear subs that he devised and helped design—never wavered. He died in 1986.

Also in August 1945 Alvin Weinberg had just moved from the University of Chicago, where he worked in the Metallurgical Laboratory devising reactors to produce the plutonium for the Fat Man bomb, to Oak Ridge, Tennessee, to help build the nation's premier lab for nuclear physics. Weinberg would become the research director of Oak Ridge National Laboratory in 1948 and director of the entire facility in 1955. Along with his mentor, Eugene Wigner, Weinberg worked out the basic design of the light-water reactor, which became the de facto standard for the world's nuclear plants. He would become one of the first public scientists in the United States, an eloquent proponent of nuclear power, and one of the harshest critics of the nuclear power industry, coining such phrases as "Faustian bargain" and "nuclear priesthood" to describe the insoluble dilemmas presented by conventional, uranium-based nuclear plants.

In the 1950s and 1960s these two men—a career naval engineer and a nuclear-physicist-turned-lab-director—helped shape the future of nuclear power in the United States and around the world. At times collaborators and ultimately adversaries, Rickover and Weinberg parted ways over the best path forward as nuclear power emerged as a major source of electricity: Weinberg, foreseeing that the first nuclear era (he used the phrase as the title of his memoir) eventually would stall, pushed for an entirely different and radical machine, the molten salt reactor (MSR), and for the use of thorium to power it. Rickover, and the men who followed him, favored conventional solid-core uranium-based light-water reactors, which as a by-product produced plutonium that could be refined for nuclear weapons.

The parallels between the two men run deeper than simple opposition, though. Both were born in the early twentieth century, Rickover in 1900 and Weinberg 15 years later. Both were the sons of immigrants from the eastern-most reaches of the Old World: Weinberg's parents were Russian, Rickover's Polish. Both grew up near Chicago and were educated at the preeminent institutions of their chosen professions: Rickover graduated from the U.S. Naval Academy in Annapolis, Weinberg from the University of Chicago. Both Jews, they were by definition outsiders in midcentury America, which had only just begun to rid itself of the anti-Semitism that had consumed European Jewry. Both transformed the organizations they came to lead and in the process inspired, promoted, and championed a generation of followers and protégés. And both were ultimately fired, dismissed by the country they'd spent their lives serving. Their entwined stories tell the history of the origins of nuclear power, and of the mistakes and missteps that got us into the energy mess in which we find ourselves now, in the second decade of the twenty-first century.

Where we find ourselves, of course, is with a stalled nuclear power industry that could provide a much larger portion of our baseload energy demand without carbon emissions. The problem is not with nuclear energy per se; it's with uranium-based nuclear reactors. How they came to dominate the emerging nuclear industry is the subject of the next three chapters. Uranium's victory was a triumph of military uses of science and technology over human-istic ones, of the Pentagon over the scientific community, bureaucracy over individual initiative, technological stasis over inspiration and innovation. It was a triumph for Rickover's worldview of an unrelenting competition be-tween implacable nuclear-armed enemies over Weinberg's vision of a peaceful and prosperous age fueled by beneficial, broadly shared nuclear technology, specifically using thorium. And it would determine the shape of today's nu-clear power industry, which has not come close to fulfilling the promise it showed at the dawn of the Atomic Age.

TH90 • TH90 • TH90

ADMIRAL RICKOVER LEFT BEHIND FEW personal accounts of his life, refusing several times to write his memoirs and tightly controlling the disposition of his pa-pers at the end of his life. An intensely impersonal man, abrasive and opaque even to those closest to him—including his long-suffering wife, whom he dragged on a series of grueling, overland journeys across prewar Southeast Asia in the mid-1930s as he moved between postings—Rickover is an elusive figure. It's hard to know what, at any moment in his eventful life, he was feeling or thinking, other than impatience and disdain for the fools who sur-rounded him.

Hyman Rickover was born in January 1900 (according to his official Navy biography and his tombstone; local school records have the date as August 24, 1898, a two-year discrepancy that a man like Rickover would not have been eager to correct) in Maków Mazowiecki in northeastern Poland. Poland at that time was under the boot heel of the Russian czar, and the Rick-overs lived in the Pale of Settlement, the region on the western fringe of the Russian empire where Jews were allowed permanent residency.[1]

Rickover's father, Abraham, was a tailor, and around the turn of the century he emigrated to New York City. A few years later—the chronology is hazy—he had scraped together enough money to invest in an apartment building in Brooklyn, and he sent for his family. Hyman, his mother, and his sister, Fanny, boarded a ship bound for the United States from one of the ports along Europe's northern seaboard, Antwerp, Bremen, or Hamburg. By 1945, four decades later, the Jewish communities of the Pale had been eradi-cated in the Holocaust.

Rickover had a hardscrabble childhood on the Lower East Side of Manhattan and, after Abraham moved the family again, in Chicago. The Rickovers lived in Lawndale on Chicago's west side, a neighborhood that would eventually be the most heavily Jewish part of the city. Rickover's intelligence and determination showed themselves early; he first went to work at age nine, making three cents an hour assisting in his neighbor's mechanic shop, and he attended John Marshall High while delivering telegrams for Western Union.

Rickover's application to the U.S. Naval Academy in 1918 succeeded only through the intervention of influential connections. His sense of being an outsider was heightened at Annapolis. Rickover was despised by his fellow midshipmen, in part because he was a Jew but more because he was a grind. Graduating 107th in his class of 540, he hardly seemed to be a future admiral when he was commissioned as an ensign on the destroyer USS *La Vallette* in 1922.

Rickover had a genius for organization, though, and it revealed itself in his early postings as engineering officer on a series of ships in the 1920s, including the USS *Nevada*. He spent a year at the Naval Postgraduate School in Annapolis, earning his master's in electrical engineering. Rickover then studied at Columbia, where he met his wife, a grad student in international law. Foreseeing the future of undersea warfare, Rickover applied for a position as an engineering officer on a submarine at the end of the decade. Again, he was denied (ostensibly because of his age—he was 29, considered too old for sub duty) until an older man intervened. C. S. Kempff, his commanding officer on the *Nevada* and by then a rear admiral, ran into Rickover outside the Bureau of Navigation in Washington, D.C., learned what happened, and interceded on Rickover's behalf. In October 1929 Rickover reported for duty aboard the submarine *S9*.

At that time submarine duty had not acquired the patina of glamour it would have in the Cold War. Subs were cramped, smelly, and greasy from diesel oil. Crews of the 1920s called them pig boats. For Rickover, though, sub duty presented a unique challenge: powered by electrical motors (which tended to fail and limited their undersea range), submarines had changed little since World War I. The young engineering officer believed he could play a role in advancing the science of underwater propulsion; he likely also saw the submarine corps as a relatively quick path to command of his own vessel.[2] His sub service lasted only four years. His only command, on the minesweeper USS *Finch,* resulted in near mutiny, as the crew chafed under Rickover's authoritarian style. After less than three months, he was relieved in October 1937; two years later he was promoted to lieutenant commander and became the assistant chief of the Electrical Section of the Navy's Bureau of Ships in Washington. He never commanded a ship again.

By December 1941, when Pearl Harbor was attacked and the United States entered World War II, Rickover was chief of the electrical section, a powerful post in a period that saw the United States create the largest and most formidable fleet in the history of warfare. Once again Rickover saw himself as battling ignorance, incompetence, and bureaucratic action in order to accomplish his aim: to create an invincible modern Navy. He became one of the most effective of the Navy's wartime officers, respected by his superiors and feared, if not admired, by his men. But he would never forget, or be allowed to forget, that he was not a front-line commander.

"Sharp-tongued Hyman Rickover spurred his men to exhaustion, ripped through red tape, drove contractors into rages," commented a *Time* magazine reporter in a postwar profile. "He went on making enemies, but by the end of the war he had won the rank of captain. He also won a reputation as a man *who gets things done.*"[3]

To Rickover, "it seemed as if he had to fight to get anything done. Rickover became mad at the Navy, and it would be an anger that would never cease."[4]

TH90 • TH90 • TH90

IF HYMAN RICKOVER HAD AN UNPARALLELED skill for making enemies and antagonizing superiors, Alvin Weinberg went in the opposite direction in his personal relations. He inspired admiration among his peers and something like love in the people who worked under him. It's impossible to find anyone who has written anything negative about him. His "special gift" was "his ability to communicate, even inspire," and his memoir, *The First Nuclear Era,* is a marvel of even-handedness: he treated gently even men who thwarted him out of personal or political motives.[5] "I have always admired Rick's courage," Weinberger wrote of Rickover, the nuclear admiral, " . . . though I did not like his autocratic methods." Those methods, Weinberg remarked, "made it impossible for people of the very highest caliber . . . to work very long with him."[6] Though Weinberg was unquestionably of the very highest caliber, he managed to work with Rickover for longer than most; unfortunately, their collaboration was not completely to the benefit of Oak Ridge, the future of nuclear power, and the country they both served.

Like Rickover, Weinberg was a product of the eastern European Jewish diaspora. Also like Rickover, he was the son of a tailor. Born in Chicago to Russian immigrant parents, Weinberg was 15 years younger than Rickover and enjoyed a smoother path to success. A graduate of Roosevelt High School, which had opened in 1922 (and which has produced a string of notables, including the writer Nelson Algren), Weinberg was only 16 when he enrolled at the University of Chicago. His first choice was Stanford, but the Great Depression had taken hold, and Weinberg lived at home during his first couple of years in college. At the university Weinberg, a gifted chemist and physicist, became part of the Chicago School of eminent physicists, many of them émigrés, who helped propel the United States to leadership of the new nuclear era—and to the building of the first atomic weapons—in the 1930s and 1940s. He also came under the influence of Alfred Korzybski, an influential psychologist who founded what became known as General Semantics, a midcentury field based on the premise that "our moods, our perceptions of each other, indeed, our mental health, are determined by the structure of our language."[7] To a degree unusual among scientists, Weinberg understood the power of language in shaping not only perceptions but public policy.

In the interwar decades, the University of Chicago had not yet become the right-wing hotbed it would during the Cold War—Weinberg later liked to joke that, at the time, everyone at the university was, if not an outright communist, at least a fellow traveler. Though he played a key role in helping build the nuclear arsenal that ultimately helped bring about the collapse of communism, his leftist tendencies—more specifically, his liberal humanism—never left him.

Weinberg's early career focused more on the life sciences than on nuclear physics. While obtaining his Ph.D. in cell biology, Weinberg gained a deep understanding of classical diffusion theory, the science that describes how particles spread and disperse through apparently random motion. Diffusion equations lie at the heart of the analysis of nuclear reactors—a subject on which Weinberg would become the world's leading expert.

Weinberg's rise to the first rank of American physicists was swift. Still in his midtwenties, he was part of the team under Enrico Fermi at Chicago that built the first nuclear reactor—or pile, as reactors were universally known at the time. Weinberg also became an honorary member of the extraordinary generation of physicists who emerged from Hungary in the early twentieth century, a group of pioneers that included Leo Szilard (who first foresaw the awesome potential of a nuclear chain reaction and patented the concept in 1934), Oskar Klein (who helped originate the idea of extra dimensions, beyond the three apparent in the physical world), John von Neumann (conceiver of the implosive lens design of the Fat Man bomb), Edward Teller (the arch anticommunist and father of the hydrogen bomb), and Weinberg's mentor, Eugene Wigner.

Wigner himself played a formative role in the history of thorium power. A native of Budapest, a 1963 Nobel Prize winner, the coauthor of the Wigner-Eckart Theorem (the earliest general theory of nuclear reactions), and the first research director of Clinton Laboratories (later Oak Ridge National Laboratory), he helped pioneer the earliest designs for thorium reactors. Wigner first met the young Weinberg in 1942 in Chicago. Weinberg was a junior scientist on what would become the Manhattan Project, and Wigner had already made important contributions in nuclear physics and quantum mechanics. The two hit if off quickly; Weinberg realized that Wigner was the most gifted theorist he would ever work with, while Wigner saw in the younger man a tireless and brilliant calculator who could give mathematical flesh to Wigner's flashes of insight.

Even before the United States entered the war on December 8, 1941, an influential group of scientists, many of whom had fled the Nazi terror, had begun thinking about and doing preliminary research on an atomic bomb. Albert Einstein's famous letter to President Franklin Roosevelt, dated August 2, 1939, and warning of the possibility that America's enemies might develop an atomic weapon, had not immediately resulted in a concerted national effort to develop nuclear weapons, but the attack on Pearl Harbor, along with the realization that Germany was almost certainly working on an atomic bomb of its own, lent urgency to the nuclear weapons research that was already underway at several institutes and universities around the country.[8] In less than a year the Army officially set up the Manhattan Project, with specific components of the bomb program assigned to designated facilities. A team at the University of California at Berkeley under Robert Oppenheimer worked on the actual design of the bomb; the study of the nuclear chain reaction and the production of plutonium and enriched uranium for the bomb's core were assigned to the Metallurgical Laboratory (Met Lab) at Chicago.

Led by Arthur Compton, winner of the 1927 Nobel Prize in physics, the Met Lab represented the most remarkable gathering of scientific minds since Niels Bohr's legendary Copenhagen conferences in the early 1930s. On hand were Fermi, Wigner (who had joined Enrico Fermi at the University of Chicago in 1942[9]), Szilard, John Wheeler, Edward Teller, Glenn Seaborg, and, of course, Weinberg. The Met Lab was both a cover name and something of an inside joke; the scientists, after all, were focused on the separation of different forms of metallic elements, but there was not a real metallurgist among them. They devised code words to refer to the top-secret materials for the bomb. "Plutonium was 'copper,' U235 was 'magnesium,' uranium, generically in the nonsensical British coinage, 'tube alloy.'"[10]

Weinberg's primary role at the Met Lab was making calculations of neutron economy—the number of neutrons produced on average for each fission event in various combinations of fissile elements and moderators—and of the appropriate size of a pile in order to sustain a chain reaction. While neutron economy is usually expressed as η by physicists, Weinberg and his colleagues in Chicago used κ. Weinberg and his team were the "custodians of κ." Although he led the effort to carry out these critical calculations, Weinberg was not among those who were invited to the signal event of the early World War II atomic effort: the first chain reaction, carried out in Fermi's "Chicago pile" (CP–1), on December 2, 1942. Decades later the omission still rankled, but Weinberg, as ever, was diplomatic about it: "I felt a little left out."[11]

Although producing plutonium was the singular goal, there were multiple theories as to the best way to get there. At first the Met Lab scientists had few certainties; one was that they had to find a way to enrich natural uranium so that the percentage of the fissile isotope U-235 rose to at least 3 percent; another was that they needed pure plutonium in quantities large enough to make a bomb. Refining or enriching uranium was one thing; the basic processes were understood. Uranium can be enriched in several ways: gaseous diffusion (where the uranium is gasified and forced through a membrane); forcing atoms through a magnetic field so that they disassociate according to mass, a method called electromagnetic isotope separation (in which charged particles traveling in a magnetic field are deflected by varying amounts that are determined by their mass) and centrifugal separation, which uses large arrays of spinning tubes in which the heavier U-238 molecules are driven toward the perimeter of the tube.

Acquiring plutonium was a challenge of a different order altogether. Identified only in 1940, the mysterious element had never been refined in more than microscopic amounts. Wigner and his team of theorists were confident that by irradiating uranium they could obtain pure plutonium; doing it on an industrial scale was a speculative endeavor at best. A large dedicated pile would have to be built. So focused on the wartime goal were the scientists of the Met Lab that few other considerations were allowed to interfere—including the notion that the piles, or reactors, the group was designing to produce plutonium for atomic bombs might one day produce vast amounts of usable energy to power civilizations, or that those power reactors would prove dangerous to operate. "We were so intent on getting something that produced plutonium with the smallest inventory of scarce uranium, that such issues simply didn't emerge."[12]

Because the Americans believed they were in a race with Germany to build an atomic weapon, they were determined to build both uranium and plutonium bombs, in case the former didn't actually work. To obtain the enriched uranium, plants for both electromagnetic separation and gaseous diffusion were built as well (later, centrifuge arrays would become the prevalent method). The plutonium production facility was located at Hanford on the Columbia River in central Washington State. The entire town was condemned and the 3,000 inhabitants relocated. But first a pilot plant was needed to prove that it could be done. The trial pile, along with the gaseous diffusion plant and an electromagnetic system for uranium enrichment, were built at a site in the hilly woodlands along the Clinch River, near Knoxville, Tennessee. Originally called the Clinton Laboratories, for the nearby town of the same name, the Tennessee facility would be one of the three primary sites of the Manhattan Project—along with Hanford and Los Alamos, New Mexico, where the bombs were actually assembled.

The first research director of the Clinton Labs was Eugene Wigner, who brought with him Alvin Weinberg and a cadre of the best nuclear scientists from the Met Lab. After the war it would become Oak Ridge National Laboratory.

TH90 • TH90 • TH90

HYMAN RICKOVER HAD A USEFUL and distinguished war, if not a glorious one in battle. He helped oversee the most massive shipbuilding program in the history of the world, and his energy, rigor, and appetite for new ideas helped shake up the traditionalist culture of the Navy. He won a Legion of Merit, the Navy's highest noncombat honor, for "exceptionally meritorious conduct . . . as Head of the Electrical Section of the Bureau of Ships." At war's end, though, he found himself effectively out of a job. The story of Rickover's rise, in less than a decade, from obscurity to one of the most powerful admirals in the history of the U.S. Navy is one of implacable personal ambition, bureaucratic knife fighting, and remarkable achievement. But it's also the story of the Navy's evolution in the first two decades of the Cold War.

The atomic bombing of Hiroshima and Nagasaki did more than demonstrate to the world that the United States now possessed a devastating new weapon, one that for a brief time made America effectively invincible in war. It also changed the very nature of war. Since the earliest Portuguese sea captains, under Prince Henry the Navigator, began exploring the western coast of Africa in the early fifteenth century, power on a global scale had meant supremacy at sea. The greatest nations had the greatest fleets. The British Empire ruled over much of the known world primarily because of its great navy. "Upon our naval supremacy stands our lives and the freedom we have guarded for nearly a thousand years," Winston Churchill declared in November 1911.[13]

This began to change in the great wars of the twentieth century, fought across broad fronts on the European landmass. But the navies of the Allies played a decisive role in both conflicts. D-Day, of course, was the largest naval invasion in history, involving five thousand ships. The turning point of World War II in the Pacific was the Battle of Midway, called by the historian John Keegan "the most stunning and decisive blow in the history of naval warfare."[14]

The Battle of Britain (the fight for air superiority over the British Isles) and the massive firebombing raids on Germany and Japan were what truly signaled the shift in the balance of war from the sea to the air. With the twin developments that resulted in Hiroshima and Nagasaki—the creation of the atomic bomb and the development of long-range bombers—that balance decisively tipped, never to shift back.

This shift was immediately understood by war planners in Washington, D.C., and Moscow. "The plane gave man great range, gave him the ability to leap over terrain barriers and seas, to pass above the struggling surface forces and to strike directly the enemy's cities, industries, communications and will to resist," wrote the military analyst Hanson W. Baldwin days after the Japanese surrender. "But the plane was a vulnerable and expensive instrument. Tremendous destructive power could be obtained only by the use of tremendous numbers, and heavy bombers, with all their appurtenances, are among the most expensive and complicated instruments of war ever invented."[15]

The dawning of the Atomic Age meant that one plane, one bomb would suffice. And the perfection of intercontinental ballistic missiles in the 1950s reduced that equation further: one missile, multiple warheads = incomparable destructive power. Advances in the technology of warfare led to an inescapable conclusion: great naval battles were tableaux from the past, and the value of naval supremacy was reduced to the fleets' ability to convey fighter planes and deliver nuclear-tipped missiles to the enemy's heart: its cities. The U.S. Navy that Hyman Rickover had known and helped to create, designed to win set-piece battles at sea, suddenly looked obsolete.

This shift brought with it not only the realignment of war-fighting priorities but also the reorganization of the U.S. Armed Forces themselves: The military aviation arm, the U.S. Army Air Forces, had since its inception been part of the Army. In 1947 it was elevated to a separate branch, the Air Force, and it commanded a growing share of strategic importance, and funding, through the early Cold War years. Overnight, The Bomb had made most military strategy obsolete, and naval strategists saw their mighty fleets, which had essentially won the war in the Pacific, becoming glorified freighters. The new age would be ruled by terror from above.[16]

This shift was not lost on the Navy's top officers, including Rickover. No new ships would be designed in the immediate postwar decade, and few new ships would be built at all. With 23 years of active service behind him,

Rickover was a temporary captain eligible for retirement. He could have moved on, like many of his contemporaries, to a lucrative second career at one of the defense contractors that were already proliferating, like parasitic organisms, in the suburbs of Washington, D.C. After the surrender of Japan, his assignment was inspector general of the Nineteenth Fleet, responsible essentially for mothballing ships. "When the ship-repair base [at Okinawa] lay in ruins and was declared superfluous to the Navy, Rickover, as commander of the ruins, seemed equally so."[17]

For a man of his restless temperament, sharp intellect, and powerful resentments, this was obviously an intolerable situation. If there was to be a war, hot or cold, he was going to be in it. Setting out to find a place for himself in the Navy of the future, Rickover soon turned his eye to the only part of the Navy that seemed likely to grow: the submarine fleet.

TH90 • TH90 • TH90

WHEN ALVIN WEINBERG CAME TO OAK RIDGE in 1945, there was little to indicate that the place would become one of the foremost nuclear energy labs in the world. "A sprawling collection of one-story wooden buildings," in Weinberg's description, it was a makeshift industrial plant dominated by the 70-foot-high black barn that housed the X-10 reactor—the pilot plant for plutonium production that had given birth to the much larger reactor complex at Hanford. The wilds of central Tennessee had not been an obvious choice. Opponents of the site noted that no great universities were nearby—the main campus of the University of Tennessee, 30 miles down the road in Knoxville, apparently didn't merit that designation—and the area had few cultural or scientific institutions, "little independent intellectual life," as Weinberg himself noted.[18]

What Oak Ridge did have was the Tennessee Valley Authority, which had been founded in 1933 and offered ample electricity, generated by dams on the Tennessee River and its tributaries, including the Clinch. David Lilienthal, the head of the TVA (and later the first chief of the Atomic Energy Commission), vocally opposed the choice of Oak Ridge. He was overruled by Arthur Compton, head of the Met Lab. The site also had the support of powerful politicians from the region and that of Eugene Wigner, for whom the hills and rivers of the area held a mysterious appeal.

"Arbitrary bureaucracy, made doubly powerful by military secrecy, had its way," Lilienthal recalled.[19] That description could apply to the entire course of the early nuclear power industry.

Spread along the valleys of the Clinch River and its tributaries, the Oak Ridge area had for generations sheltered small hill communities like Wheat, Elza, and Scarborough. Farmers along the Appalachian foothills still plowed with mules, and electricity was a relatively recent innovation. Hundreds of families were relocated, their property condemned, after the Army Corps of Engineers took over the site in 1942. In less than two years, at a cost of about $1.2 billion, three of the major operating plants for the Manhattan Project were carved from the forest. To the west was the gaseous diffusion plant, K-25, which covered more ground than any structure in history. South rose Y-12, where U-235 was separated using electromagnetism, and southwest was X-10, built in nine months, the first large functioning nuclear reactor in the world.

For the scientists and their families, used to the urban delights of Chicago, life at Clinton Labs was spartan. Most employees lived in military-style barracks; more comfortable brick homes for families were constructed only after the war. In the early days, a borrowed circus tent served as the cafeteria and an old schoolhouse as both office space and dormitory. In an early instance of the blending of private-sector technology with military aims, operation of the facility was turned over to the chemical giant DuPont, whose executives continually butted heads with Wigner during the war and the first years of peace.

Weinberg brought his wife, Margaret, and their two-year-old son Richard on the Louisville & Nashville train from Chicago to Knoxville. They arrived in May 1945, and he lived and worked in the area for most of the rest of his life.

Charged with providing the basic fissile material for the atomic bombs detonated at the Trinity test site in New Mexico, and subsequently dropped on Hiroshima and Nagasaki, the Clinton Labs had succeeded brilliantly. But at war's end, Weinberg, like Rickover, found himself attached to a beleaguered institution—one that, unlike the Navy, had to justify not only its continued preeminence but its very existence. The X-10 reactor and the uranium separation plants had been constructed for one purpose: to make possible the atomic bombs that ended the war. What would be their role in postwar America?

In the days after the war, "staff members drifted about Clinton Laboratories, gathering and talking, seemingly bereft of energy. 'Everyone,' admitted one scientist, 'felt a sense of disorientation, of slackness, of loss of direction.'"[20]

Responsibility for the nation's nuclear arsenal had been shifted to Los Alamos, Hanford, and Argonne, outside Chicago. Generating electricity from nuclear fission was still just an idea, one that many experts believed would never be economically feasible. Much of the postwar military leadership, including Leslie Groves, military head of the Manhattan Project, believed that Clinton should simply be shut down. They failed to reckon with the energy and vision of Eugene Wigner.

As early as 1944, Wigner—who followed his disciple Weinberg to Tennessee in 1946—had drawn up a plan for an extensive postwar nuclear research facility, staffed by as many as 3,500 people and dedicated to devising new, peaceful applications for the fearsome power that had devastated two Japanese cities. As one of the architects of the Manhattan Project, Wigner had unquestioned status with Congress and the White House, as well as a keen sense of the fears and desires of American politicians. Science and industrial might, along with the dogged courage of the troops, had won the war for the Allies. Preserving the nation's technological preeminence was a primary goal of the postwar leadership, which foresaw that the secrets of building the bomb would not remain exclusive indefinitely (though predictions of how soon the Soviet Union could build its own bomb proved to be off by a decade or more). A system of national laboratories seemed like a good hedge against future nuclear competitors. While at this time the concept of a nuclear power reactor had yet to be proven (and not everyone believed that it would ever be), the scientists who had harnessed nuclear power for unprecedented destruction—and who, faced with the carnage in Hiroshima and Nagasaki, were already haunted by remorse—had a powerful urge to harness the atom for peaceful purposes.

Future reactor development had already been taken from the Oak Ridge scientists—stolen from them, in their view, and reassigned to Los Alamos and Argonne. Weinberg refused to believe that Clinton had no role to play in nuclear power going forward; the home of reactor technology for the Manhattan Project, Wigner and Weinberg argued, should become a center for advanced research into chemical metallurgy, radioisotopes, and nuclear chemistry. And it could become the primary training ground for future nuclear physicists. In 1949 Oak Ridge National Laboratory was created, along with its associated graduate school, the Oak Ridge School of Research Technology—the Clinch College for Nuclear Knowledge, the scientific wags called it.

In the next decade the school would provide instruction into the atom's mysteries to a generation of postgrad physicists, visiting students from friendly foreign countries, and a succession of military officers, including Hyman Rickover.

TH90 • TH90 • TH90

THE SECOND HALF OF THE 1940S was an odd interregnum in U.S. history, a brief, sunny period of hope and optimism soon darkened by the growing clouds of the Cold War.

In the late 1940s the rush of wartime science and technology research gave way to advances that would benefit humanity instead of killing men and women. At Bell Labs in New Jersey, the Solid State Physics Group under William Shockley developed the solid state transistor, leading to the semiconductors that would fuel the information revolution. In 1947 Chuck Yeager broke the sound barrier in the experimental aircraft Bell X-1, and the Russian defector George Gamow carried out the calculations that explained the big bang theory of the origins of the universe.[21]

And at Oak Ridge, Tennessee, Wigner and his team explored the future of nuclear reactors. Even before the war's end, having accomplished their main Manhattan Project task, the proof-of-concept for the Hanford plutonium-producing reactors, the scientists had time to think about the future of nuclear power. The New Piles Committee was formed to develop ideas. It considered many different types of piles, including an "aqueous homogeneous" system that could bombard a blanket of thorium-232 to create fissile U-233 in its core, dreamed up two years earlier by Eugene Wigner.

"Crazy ideas and not-so-crazy ideas bubbled up," Weinberg later wrote, "as much as anything because the whole territory was unexplored—we were like children in a toy factory."[22]

They were working in a time of great political ferment as well. There was a moment after the war when it seemed that leaders in the East and the West might avert a nuclear weapons competition rather than plunge the world into a nightmarish arms race that carried the threat of mutually assured destruction. Haunted by the possibility of annihilation unveiled at Hiroshima and Nagasaki, a group of scientists and senior figures in the Truman administration—and, for a time, the president himself—argued forcefully for freely sharing the secrets of nuclear energy with the communist adversary. In a series of memos to the president, Secretary of War Henry Stimson outlined the fateful choices facing the only nation to have exploded a nuclear weapon. Nuclear power, Stimson wrote, "caps the climax of the race between man's growing technical power for destructiveness and his psychological power of self-control and group control—his moral power." The question of sharing nuclear technology "becomes a primary question of our foreign relations," Stimson declared, adding that a proposal should be made to Moscow "just as soon as our immediate political considerations make it appropriate."[23]

Truman's "official policy initiatives through 1948 focused exclusively on the goals of establishing civilian control over American nuclear resources . . . and seeking international control of atomic energy in the United Nations."[24]

That moment passed. In an address to Congress in March 1947, President Harry Truman laid out what would become known as the Truman Doctrine: it must be "the policy of the United States to support free peoples who are resisting attempted subjugation by armed minorities or by outside pressures." The world was now divided into two camps at perpetual war with each other, and, as expressed in a document written by Truman's aide Clark Clifford and clearly supported by the president, "the United States must be prepared to wage atomic and biological warfare."[25]

Two years later, about a decade earlier than U.S. experts had predicted, the USSR tested its first atomic bomb, and the nuclear arms race was on. Born in the fires of World War II and proven in the utter destruction of two ancient cities, nuclear power was now the primary instrument of an implacable conflict between the two most powerful nations the world had known. The prospects for the peaceful development of clean, safe nuclear power were eliminated. Uranium, and the bomb, had won.

TH90 • TH90 • TH90

IN OAK RIDGE, MEANWHILE, the euphoria of victory in war gave way quickly to the disillusionment and disarray of peace. Eugene Wigner returned to his post at Princeton in 1947, and several prominent scientists turned down the position of laboratory director. Once the source of much of the science that fueled the Manhattan Project, Oak Ridge became a backwater. The Atomic Energy Commission (AEC), formed in 1946 with former TVA president David Lilienthal as its first chair, had designated Los Alamos as the nation's primary weapons laboratory and Argonne as the site for all reactor work. After Wigner's departure almost a year went by with no official director of the Clinton Lab. No one wanted to take over a place that seemed "destined for extinction."[26]

Inevitably scientists began leaving for more attractive posts at the other national labs and at universities eager to build up their physics departments. When the AEC named the Carbide and Carbon Chemicals Company, which had run the gaseous diffusion and electromagnetic enrichment plants at Oak Ridge during the war, to take over management of the entire laboratory, "all hell broke loose," and the lab faced a mass exodus. Renamed Oak Ridge National Laboratory (ORNL) in 1948, the place was caught in a downward spiral of controversial decisions, questionable management, and an uncertain mission.

Weinberg remained characteristically steadfast. He didn't share his colleagues' mistrust of the Carbide company, and he had come to love the area. He set about trying to broker a truce between the researchers and the managers, and in March 1948 he was asked to take over. Wary of the administrative duties and the political land mines of the director's office, Weinberg agreed instead to become associate director. He would become the research director a year later and full director in 1955, and he essentially defined the laboratory and its mission for the next quarter century.

Weinberg believed that Oak Ridge had a future as a center for nuclear reactor research: "I didn't think that the decision to concentrate all reactor development at Argonne could be carried out."[27] What's more, he had in mind a specific mission for ORNL: building a homogeneous (liquid fuel) breeder reactor based on thorium and its fission offspring, U-233. Weinberg had come to believe that liquid fuel thorium reactors would transform the nation's energy supply. He spent the next 20 years attempting to bring that vision to reality. The story of his failure is told in chapter 5; here I will examine the forces that combined to shape the nuclear power industry in the Pentagon's image.

In 1945–46 Weinberg led a campaign to institute civilian control over the new AEC, lobbying for a bill sponsored by Senator Brien McMahon, a Democrat from Connecticut, that would establish a five-member civilian commission and a general manager who was also a civilian. The Pentagon lobbied hard for the competing May-Johnson bill, which would effectively keep development of nuclear power under military control. The scientists won; the McMahon bill passed, but it was an illusory victory. After pressure from conservatives forced several major revisions to the bill to appease the military, Truman signed it as the Atomic Energy Act of 1946. The act did not "in any respect diminish the dominance of the military in nuclear affairs," Weinberg would write.[28]

With the explosion of the first Soviet nuclear weapon at Semipalatinsk in Kazakhstan in August 1949—almost precisely four years after Hiroshima—that dominance reached into other parts of American industry, commerce, and culture. "The Pentagon was the center of an implicit and . . . profound militarization of society" that combined "economic, political, martial, academic, scientific, technological and culture forces," James Carroll wrote.[29] To men like Edward Teller, Air Force Chief of Staff Curtis LeMay, and George Kennan, a chief architect of the hawkish U.S. foreign policy based on supposed supremacy in nuclear weapons, it was inconceivable that the primary object of this ultimate power source could be anything but war. Civilian nuclear power was a distant second; and the makeup, goals, and decisions of the AEC all would reflect this. David Lilienthal was hardly a doctrinaire cold warrior, but the commission he chaired had no illusions about the real identity of its masters. "The requirements of national defense thus quickly obscured the original goal of developing the full potential of the peaceful atom," wrote Alice Buck in an official history of the AEC commissioned by the Department of Energy in 1983. "For two decades military-related programs would command the lion's share of the Commission's time and the major portion of the budget."[30]

National defense requirements imposed three basic limitations on Weinberg and the others who sought to develop a peacetime nuclear power base: All scientific data relating to nuclear technology was classified, severely restricting information flow. Innovation in nuclear power was subservient to the maintenance of superiority in the arms race; of premier importance was ensuring a sufficient supply of weapons-grade uranium and plutonium at a time when the world's available reserves of uranium were believed to be scarce. Finally, reactor development, in both the short and long term, was channeled into programs that would directly benefit military operations—meaning, in the first case, submarine propulsion.

Those limitations would eventually doom Weinberg's dream of a thorium-based homogeneous breeder. But they would pave a broad way for the dream Hyman Rickover brought with him to Oak Ridge: a nuclear-powered submarine armed with atomic bombs. The ultimate attack vehicle: the USS *Nautilus*.

FIVE

THE BIRTH OF NUCLEAR POWER

I n March 1946 a young physicist from the Naval Research Laboratory named Philip Abelson, the codiscoverer of the element neptunium and a primary developer of a process to separate U-235, submitted a report to the Navy brass entitled "Atomic Energy Submarines."

Work at the Naval Research Laboratory had indicated, Abelson wrote, that "only about two years would be required to put into operation an atomic-powered submarine mechanically capable of operating at 26 knots to 30 knots submerged for many years without surfacing or refueling." Though much of the report was later considered "vague or—from a technical viewpoint—even questionable," it galvanized support for a nuclear sub program and would help shape nuclear reactor technology for the next four decades.[1]

Abelson submitted his findings to a group of about three dozen high-ranking officers later that month. "If I live to be a hundred," Rear Admiral Charles Lockwood, one of those present, recalled later, "I shall never forget that meeting . . . in a large Bureau of Ships conference room, its walls lined with blackboards which, in turn, were covered by diagrams, blueprints, figures, and equations which Phil used to illustrate various points. . . . It sounded like something out of Jules Verne's *Twenty Thousand Leagues Under the Sea*."[2]

Not science fiction but the brute realities of the arms race led to the meeting. At least in the minds of strategic planners in the United States, superpower warfare had been reduced to a simple calculation, complicated in the execution: how to deliver nuclear warheads most effectively to destroy the major cities of the Soviet Union without being detected. Bombers could be seen by radar and shot down; submarines could slip undetected to within miles of the enemy's coastline to deliver Armageddon. First-strike submarines that could remain submerged, and run silently, for months or even years at a time became the absolute strategic weapon of national security policy and the primary goal of the postwar Navy.

Quickly it was decided that a group of five naval officers (along with three civilians) would travel to Oak Ridge to observe the nuclear reactor R&D work under Wigner and Weinberg. The officers were ordered to Tennessee in the summer of 1946. Captain Hyman Rickover was one of them.

Rickover arrived in June 1946—as the lab was descending into chaos and near-mutiny. He was the most senior of the naval officers, who became known, ironically or not, as "doctors of pile engineering," or DOPEs. Edward Teller, the father of the hydrogen bomb, would later write of his introduction to the "unknown Navy captain," saying that Rickover shook his hand and said, "I am Captain Rickover. I am stupid."[3]

While no match for the brilliant scientists at Oak Ridge, Rickover was hardly stupid. His genius lay in organization and systems analysis. At 46 he was a bantamweight terror: just 5'6" and 125 pounds, he stared up at more sizable commanders; his sharp features and piercing eyes gave him, at times, the look of a startled sparrowhawk. His career by this point had taken on a whack-a-mole quality: pushed aside or thwarted in one place, he always managed to rise in another. Much later, when eight nuclear subs had been launched and Congress had voted him the first of two Congressional Gold Medals, the *Life* magazine writer Robert Wallace would refer to "the series of official snubs that have marked his career since he emerged to prominence as the 'father' of the atomic submarine."[4]

Rickover quickly sized up the situation at Oak Ridge: left to their own devices, the Manhattan Project scientists would take their own time to develop nuclear propulsion. More important, the Navy, in the person of Hyman Rickover, would not control the technology. With little thought for such practical concerns as steam generation or actual electricity, "the physicists spun elegant, complex designs with liquid metal coolants, magnetohydrodynamic pumps, and plutonium-fueled cores using fast-neutron fission."[5]

This theoretical design work was unacceptable. Rickover wanted a nu-clear power plant now, and he wanted it for the Navy. He famously remarked that building a power reactor was "95 percent engineering and 5 percent nuclear physics." He set about transferring nuclear knowledge to his officers, driving them as relentlessly as he drove himself and steamrolling all objec-tions and hesitations on the part of his superiors and their nominal bosses, the politicians.

Rickover had his men attend every lecture at Oak Ridge. Like Kirk So-rensen would do in the early 2000s, Rickover spent his evenings poring over tomes on nuclear physics and reactor engineering, effectively adding a mas-ter's in nuclear physics to his electrical engineering degree. "He expounded and orated, needled and wheedled, and in the end was recalled from Oak Ridge, assigned some vague 'advisory duties' and given an office in an aban-doned ladies' room in the Navy Building."[6] Somehow, he got his plan for a nuclear sub into the hands of the Navy's chief of operations, Admiral Chester Nimitz. In short order, Rickover was rescued from the former women's loo and put in charge of the newly created Nuclear Power Division of the Bureau of Ships. Then, in a remarkable coup de force, he convinced Lilienthal to set up a naval reactors branch at the Atomic Energy Commission (AEC)—and had himself appointed its head. Now, "Rickover, wearing his captain's hat would write letters to be opened by Rickover, wearing his civilian hat. What-ever Rickover wanted, Rickover got."[7]

Rickover had his program. He had a talented, handpicked bunch of of-ficers almost as driven as he was. He had near carte blanche to obtain materiel and resources, and he had unquestioned authority to see his orders carried out. Now all he needed was a reactor.

TH90 · TH90 · TH90

BY THE TIME RICKOVER AND HIS MEN ARRIVED, "our focus" at Oak Ridge, Weinberg said, "was a bit confused." At the time there were three main streams of re-actor research and development, distinguished by their forms of coolant: sodium cooled, gas cooled, and water cooled. One of the ironies of Wein-berg's long career is that, through his association with Rickover and the naval nuclear propulsion program, Weinberg became the primary progenitor of the light-water reactor—which would become the dominant reactor type in the world, effectively overriding alternative technologies, including liquid-core, thorium-fueled reactors.

Weinberg, who'd become director of Oak Ridge almost by default, was just coming into his role as the leader in the new era at the lab. A born mediator, he spent much of his time making peace between the scientists and their corporate superiors. He had an expansive forehead, with mirthful eyes and long, dangling earlobes; as he aged, he started to resemble a laughing Buddha (but with hair). His intellect in some ways was the opposite of Rickover's. Broad minded, objective, and able to view multiple perspectives, Weinberg had the quality that F. Scott Fitzgerald identified as the mark of a first-rate intelligence: "The ability to hold two opposing ideas in mind at the same time." In the bureaucratic battles of the next two decades, Weinberg's open-mindedness would not always serve his purposes well.

In 1944, in a series of experiments at the Clinton Lab, Weinberg and his colleagues had realized, almost by accident, that a simple mixture of uranium and water would come tantalizingly close to chain-reacting, meaning that a nuclear power reactor could use slightly enriched uranium as the fuel and ordinary water both as moderator and coolant. In September 1944, while the Allies marched toward Berlin and the U.S. fleet grappled with the Japanese in the western Pacific, Weinberg sent a memo to Richard Doan, the research director at Clinton, explaining his ideas about a pile cooled with ordinary—that is, light—water. The primary problem, as Weinberg saw it at the time, was developing a coating (or cladding) for the uranium fuel elements that would withstand corrosion: "The advantages of a system moderated with water are obvious. Such a system could contain within itself a means for cooling . . . [and] would probably be much more compact and consequently simpler to build than the conventional piles. Finally, if the coating problem can be adequately solved (by the use, say, of beryllium) it may be possible to run such a system under pressure and obtain high-pressure steam which could be used for power production."[8]

This was the first official mention of what would become the light-water reactor. Two years later Weinberg wrote a longer analysis with a Clinton mathematician named Forrest Murray, "High Pressure Water as a Heat Transfer Medium in Nuclear Power Plants." The system they described was not a uranium-fueled reactor; it was a thorium breeder, similar to the one that was eventually built at Shippingport, Pennsylvania, ten years later. At the time Weinberg did not understand what he'd wrought. It was still possible to believe in a world where more than one reactor technology would be developed and competition would determine the optimal machine. "We had no idea that the pressurized-water reactor (PWR) would become the primary commercial power reactor," he wrote.[9]

That it did so was largely thanks to Captain Rickover, and to the exigencies of the death struggle now being waged between the United States and the Soviet Union. "Production of plutonium and U235, and later of tritium (for the hydrogen bomb), were the driving forces of the nuclear effort," Weinberg remarked.[10]

Rickover believed that the type of reactor was a secondary consideration to the overall design of the submarine, the space limitations on board, and the power required to turn the propeller at the desired speed. He had a particular design in mind: the German Type XXVI subs, which had been discovered partially assembled at Axis shipyards after the fall of Berlin. Designed to move through the water quickly and quietly, the Type XXVI had a hull width of 28 feet. This became the model for the *Nautilus* (the actual reactor compartment wound up being 27 feet, 8 inches in diameter), and "this measurement would have a profound effect on the far future of nuclear power."[11]

At first Rickover favored liquid sodium as the coolant for his submarine reactor. He knew the liquid sodium reactor program at GE in Schenectady, New York, was well advanced, and he was in a hurry. Sodium had some clear advantages: It has a high heat capacity, which makes it useful for building compact reactors. At atmospheric pressure, it has an extremely high boiling point of 880 degrees centigrade, eliminating the need for a pressurized vessel. It is a poor moderator, which is both a benefit and a disadvantage: it keeps the energy level of the neutrons, and thus the reactivity of the core, high, but it allows for less controllable reactions. And it is not corrosive, meaning that no exotic materials were needed for the piping and pumps.

Sodium, however, has one obvious flaw, particularly for use in a marine environment: it reacts explosively with water. Were the liquid sodium coolant in a nuclear submarine power plant to come into contact with water at any point, an explosion and fire would result. To his credit, Rickover was quickly persuaded when Alvin Weinberg presented the concept of a water-cooled reactor. Pressurized water had two main advantages: like a sodium-cooled system, a water-cooled reactor could be built small enough to fit with a Type XXVI–sized hull, and, as Weinberg put it, "water, unlike sodium, was something the Navy ought to know about."

Put another way, the Navy had the best plumbers in the world. They knew how to design and operate pumps, bearings, and valves to transport water, including water at the high pressure required for a nuclear reactor inside a submarine.

So the pressurized water reactor (known as the PWR or LWR, for light-water reactor) was born—"not as a commercial power plant, and not because it was cheap or inherently safer than other reactors, but rather because it was compact and simple and lent itself to naval propulsion," Weinberg wrote. "But once pressurized water was developed by the Navy, this system achieved dominance for central station power."[12]

From a distance of more than five decades, the irony of this development is evident. Built for the relatively tight spaces and specialized needs of a submarine, the LWR would power cities, becoming the dominant technology for an industry that supplies nearly 20 percent of the world's power. And the essential element of that technology—its use of pressurized water—has become its chief Achilles heel. The most common cause of nuclear accidents is not a problem with the radioactive core itself (i.e., a meltdown) but an explosion of radioactive steam, as happened at Chernobyl in 1986. Even the best plumbers have not been able to fully master the intricacies of pressurized water in a nuclear reactor.

Which makes it doubly ironic, since at the same time he was convincing Hyman Rickover to build the *Nautilus* around a pressurized-water reactor, Alvin Weinberg was becoming more and more fascinated by a reactor design that required no high-pressure vessels: the molten salt reactor (MSR).

TH90 • TH90 • TH90

WHILE THE BIRTH OF THE ATOMIC AGE is usually traced to the successful chain reaction in the Chicago pile in December 1942, the world's first successful reactor actually had a liquid core. In 1940, with Britain fighting for its life, scientists at the famed Cavendish Laboratory in Cambridge built a reactor with a core of 112 liters of heavy water mixed with powdered uranium-308. This radioactive slurry, as they called it, was contained in an aluminum sphere about 60 cm in diameter, or the size of a beach ball. A very heavy, very nasty beach ball. The sphere was suspended in a bath of mineral oil, which reflected neutrons back into it. Ordinary water didn't work: only slurry containing heavy water would sustain the reaction. Heavy water is chemically similar to plain old tap water except that the paired hydrogen atoms, as in H_2O, are replaced by deuterium atoms, becoming D_2O. Deuterium is an isotope of hydrogen with one extra neutron. An effective moderator of fissioned neutrons, H_2O also tends to absorb them, which is why light-water reactors require enriched uranium to sustain the reaction. Heavy-water reactors, such as those developed in Canada, can run on natural uranium. The pursuit of sources of heavy water, then as scarce as uranium, became one of the more obscure and compelling subplots of World War II and was the basis of the 1965 Kirk Douglas movie *The Heroes of Telemark*.

When the British scientists tested the slurry, they discovered evidence of a chain reaction. It wasn't much, barely enough to power a lightbulb, but it was fission. Liquid-core reactors thus predated solid-core ones.

How much attention Eugene Wigner paid to the Cavendish sphere is not clear, but by 1943, having completed the basic design of the plutonium-producing pile at Hanford, he became intrigued by the concept of an aqueous homogeneous (AH) pile: homogeneous because the core elements of the reactor would be in the same physical state; aqueous because that state would be water based. Experiments went forward to find a solution of uranium and water that would be stable and capable of sustaining a chain reaction. The advantages were obvious: the design of a homogeneous reactor, essentially a metal sphere, would be far simpler than the solid fuel pile, which at that stage was a complex lattice of graphite and uranium in bricks. The chemists at Clinton Lab saw the beauty of making the chain reaction essentially a chemical one. The homogeneous reactor would not have to be regularly powered down to extract spent fuel and remove xenon-135. "Gone would be the thousands of carefully machined uranium slugs," wrote Weinberg; "gone, too the intricate system of piping required to feed water into each process tube."[13]

The primary problem was the classic plumbing problem: corrosion. With hydrogen and oxygen combined with heat and radiation in a fluid core, metal corrosion was a serious challenge, and, in fact, given the metallurgy of the time, it prevented the AH pile from being used during the war. But Wigner, after all, was originally trained as a chemical engineer. He believed that the corrosion problem could be solved, and his fierce belief in his own ideas inspired his protégé, Weinberg, who later wrote: "I myself became bitten with the 'homogenous' bug, a fixation I have never recovered from!"[14]

As it happened, the first aqueous homogeneous reactor on U.S. soil—and the third fully operational reactor ever—was actually built at Los Alamos under the direction of Enrico Fermi. Code-named Water Boiler, it used the nation's entire supply of U-235 at the time. It was also called the LOPO, for low power, because, like the Cavendish ball, it produced almost zero power. It was a stainless steel sphere containing a solution of enriched uranyl sulfate and surrounded by blocks of beryllium oxide to reflect neutrons. After achieving criticality in May 1944, with Fermi at the controls as he had been for the Chicago pile two years earlier, LOPO was dismantled a short time later to make way for a second boiler—this one with a slurry of 560 grams of enriched uranium nitrate dissolved in about 14 liters of light water. Called HYPO, the new AH reactor actually produced power, 5.5 kilowatts' worth. The following year a young chemist at Clinton named Harrison Brown proposed a larger version of the water boiler that would put out ten times as much heat. Excited, Weinberg led investigations into a prototype that would produce not only power but new fuel: an aqueous homogeneous breeder using not uranium slurry but thorium and its daughter product, U-233. It was the first proposal for what, today, has evolved into the liquid fluoride thorium reactor.

Weinberg and his men were stymied by what they didn't know. Uranium was still thought to be a scarce element (a conclusion, by the way, that Rickover strongly disagreed with—he would be proved right). That meant they had to understand the fission properties of U-233, specifically its neutron flux, much better than they did at that point. That is, they had to know the number of neutrons produced by fission when a U-233 nucleus absorbs a neutron. Unless that number was comfortably greater than two, a thermal liquid-core breeder based on the thorium–U-233 fuel cycle was impossible. The thorium-based aqueous homogeneous reactor was shelved for the time being in favor of a heterogeneous reactor fueled with highly enriched

U-235—the "high-flux pile," which would become known as the materials testing reactor.

Soon, however, the idea of a fluid-fuel reactor arose again, in one of the most outlandish research programs ever carried out by U.S. scientists: the effort to build and fly airplanes powered by nuclear engines.

TH90 • TH90 • TH90

NOT TO BE OUTDONE BY THEIR RIVALS in the Navy, the generals of the U.S. Army Air Forces, eager to establish their force as a full-fledged branch of the military after the war, embraced the notion of a strategic nuclear bomber that could stay aloft indefinitely without refueling. In a time of boundless atomic mania, the answer seemed obvious: a nuclear engine.

During a congressional hearing in October 1945, a skeptical senator Homer Ferguson, a Michigan Republican, posed the question: "Do you see a future for atomic power in an airplane?"

J. Carlton Ward Jr., president of Fairchild Engine & Airplane Corporation, a leading military supplier, responded confidently: "The whole tactical concept of war will change to the nation that first solves that problem."

The next day the *Chicago Tribune* blared a front-page headline: "Predicts Atom Will End Limit on Plane Range."[15]

Never mind that the shift in tactics to nuclear submarines was already well underway or that the advent of intercontinental ballistic missiles (prototypes of which had been tested in Nazi Germany only months before the fall of Berlin) would soon further obliterate the utility of massive bombers; the Air Force high command wanted a nuclear aircraft and, by God, they were going to get it. A study at the Applied Physics Laboratory at Johns Hopkins University found, sure enough, that a nuclear-powered aircraft was a good idea. Funding was obtained, and the Fairchild company, headquartered just five miles from Oak Ridge, plowed ahead with the idea. Soon Weinberg and his team found themselves working on the Nuclear Energy for Propulsion of Aircraft (NEPA) program. Eventually NEPA would involve fourteen thousand people in seven states and consume at least a billion dollars.

The experts were unimpressed. Robert Oppenheimer called the idea hogwash. Edward Teller, who met few weapons systems he didn't support, called it "too dangerous." Weinberg himself would later write, "A nuclear aircraft was an oxymoron."[16] There were several reasons for this. For one thing, weight matters in a plane, and nuclear reactors are not light. Submarines, regardless of how many tons they weigh, can be made buoyant. An airplane needs a certain ratio of thrust to weight. Putting a reactor on board pushed that ratio beyond reasonable limits.

Second, while in a submarine it's relatively simple to shield the crew from radiation, it's not so easy in a plane, even a big bomber the size of half a football field. Shielding adds weight—around 12.5 tons in the favored configuration. At one point some Air Force genius actually proposed using older men as crews for the atomic bombers, so that the radiation dosage in their cells would have less time to accumulate during their careers.[17]

The worst problem, though, was what would happen if the plane went down. A failed nuclear submarine sinks; the crew is either lost (as in the case of the USS *Thresher,* which sank with all hands on April 10, 1963) or rescued, but the chances of further damage to the surroundings are minimal. The sea floor is actually an ideal place to store a decommissioned nuclear reactor. If a nuclear aircraft went down, it would spew radioactive debris across a swathe of territory five miles or more long and a mile wide. A failed mission would likely cause as much or more destruction than a successful bombing.

As armaments programs often do, however, NEPA gathered its own momentum. Backed by both the Air Force (which achieved separate branch status in September 1947, under the National Security Act of 1947) and the AEC, the project was recast as the Aircraft Nuclear Propulsion (ANP) program. Despite a long history of "competing egos, crossed purposes, differing priorities, and general chaos," an experimental atomic plane, the NB-36H, flew 43 test missions in the late 1950s that were deemed successful.[18] *Success,* in this case, meant the plane avoided a disastrous crash.

Still, no one could actually convince Congress that these atomic behemoths were worthy of production. President John F. Kennedy finally canceled the ANP in March 1961, just two months before he announced the Apollo program to land a manned spacecraft on the moon. The atomic folly of ANP had proven the scientists right: nuclear-powered flight was too dangerous, too expensive, and too cumbersome. But one positive development did come out of the ANP effort: it led to arguably the crowning achievement of Alvin Weinberg's long career: the Molten Salt Reactor Experiment.

TH90 • TH90 • TH90

NONE OF THE SCIENTISTS AT OAK RIDGE would ever admit that ANP was a make-work program, a sort of New Deal for nuclear physicists and engineers, but that's what it amounted to. In 1947 Oak Ridge had essentially been tossed out of reactor development; now, as the designated lead site for the aircraft reactor program, it was back in. "It wasn't that I had suddenly become converted to a belief in nuclear airplanes," Weinberg wrote. "It was rather that this was the only avenue open to ORNL for continuing in reactor development. That the purpose was unattainable, if not foolish, was not so important."[19]

He made the most of the opportunity. Many materials and technologies developed for the aircraft reactor experiment had applications for the MSR. To generate enough power to fly a bomber, the aircraft reactor had to operate at red heat; thus much of the work on ANP was aimed at developing metals and ceramics that could withstand such high temperatures. The original concept for the aircraft reactor was a liquid metal–cooled indirect cycle that used many long, thin cylindrical fuel elements placed inside a block of beryllium oxide, with liquid sodium or another liquid metal flowing over the elements and delivering heat to an exchanger that would heat compressed air to drive the jet engines. This design had all the complexity of liquid metal–cooled designs with few of the advantages. Weinberg assigned Ray Briant, a brilliant chemical engineer and applied mathematician, to head the ANP program. Though he came from the Applied Physics Lab at Johns Hopkins, Briant was no fan of nuclear aircraft, even less of the liquid sodium design. He believed there was no way the skinny, jackstraw-like fuel elements could stand up to temperatures of 1,500 degrees Fahrenheit or higher. Briant swiftly arrived at a proposal for the ANP that was sure to win Weinberg's enthusiasm: a fluid-fueled reactor. The problem was finding a uranium-bearing liquid that would be stable at 1,500 degrees Fahrenheit when in contact with the metallic pipes that would contain the liquid.[20]

One possibility was molten sodium hydroxide (NaOH), or caustic soda, commonly used as a chemical base in the manufacture of textiles, pulp and paper products, soaps and detergents, even bottled drinking water. But the corrosion problems with NaOH seemed insurmountable. A pair of Oak Ridge chemists suggested another option: molten fluorides of metals such as lithium, sodium, or potassium. Because of the strong positive charge of the alkali metals and the strong negative charge of the fluorine, these would be among the most stable compounds on Earth. If oxygen were prevented from entering the system, Briant surmised, the fluorides would not attack steel alloys, even at high temperatures—thus potentially solving the corrosion problem. To be sure, there were challenges: a high-temperature molten fluoride reactor would require temperature and radiation sensors that worked under intensely radioactive conditions; valves that would still open and close in the molten fluoride environment; and some way to measure the chemical state of the reactor fuel. But these were chemical engineering problems, ones that Weinberg and his team were both uniquely qualified and eager to solve.

The Aircraft Reactor Experiment (ARE), built in the old Y-12 electromagnetic separation building at Oak Ridge with a fluid fuel of uranium fluoride mixed with fluorides of sodium and zirconium ($NaF-ZrF_4UF_4$), went critical on November 3, 1954. "As a chain reactor," Weinberg exulted, "it behaved beautifully." Because of the high negative temperature coefficient of the molten fluoride fuel, the reactor was self-adjusting: its power output automatically adjusted to the power demand of the heat exchanger, which was determined by the amount of air flowing through the exchanger. The ARE ran for 100 hours before being shut down by Air Force Colonel Clyde Gasser in a small ceremony. About four hours after shutdown, a pipe leading to the core ruptured, spilling fission fragments into the Y-12 building, which had to be evacuated. The radiation returned to background levels within a few hours, and the Oak Ridge scientists were delighted with the experiment.

Ray Briant did not live to see the ARE in action; he died of lymphoma a few months before the reactor went critical. Weinberg, however, grasped strongly the possibilities of the molten fluoride reactor; in particular, he was convinced that a thorium-based liquid fuel reactor could be a breeder, producing more fuel than it consumed. The ANP was the direct forebear of the molten salt breeder. As I will discuss, the development of breeder reactors was the primary goal of the AEC in the 1950s. And that goal was based on a mistaken assumption: that uranium was a scarce element.

TH90 • TH90 • TH90

THE QUESTION OF HOW MUCH URANIUM there is in the world was thoroughly investigated between 1932 and 1950 by the world's best scientists. And almost all got it wrong. In 1942, at the outset of the Manhattan Project, it was believed that the world's total reserves of exploitable uranium might be no more than a few thousand tons. Seeking a reliable supply for the Bomb, Captain Groves's men secured a contract with the Union Miniére de Haut Katanga, owner of the world's richest uranium mine, at Shinkolobwe in the Belgian Congo. After the war both the United States and the Soviet Union undertook extensive uranium exploration programs; extensive deposits would be found in the American Southwest on the Colorado Plateau, and in Siberia. But the prospects for a future nuclear power industry, in the view of most scientists and policy makers, lay in securing supplies of uranium that could be exhausted quickly. The solution, as I have detailed, was to create a breeder reactor.

In the mid-1950s there were essentially two main R&D avenues for breeder-reactor development: liquid metal–cooled machines and everything else. With the ARE at an end, cognizant that nuclear-powered aircraft were a dead end, and acutely aware that Oak Ridge needed a new field in which it could establish a premier research program, Weinberg saw the lab's future in molten salt breeders. It was to prove a nearly 20-year struggle that would define his tenure as director of Oak Ridge.

"All of the other materials and coolants being suggested for [breeder] reactors had been anticipated by the reactor design group at the Metallurgical Laboratory in Chicago during World War II," wrote H. G. "Mac" MacPherson, a Union Carbide chemist whom Weinberg hired as the program director for the molten salt program. "This was new."[21]

As I've discussed, this statement, while specifically true, is conceptually misleading: Eugene Wigner had envisioned aqueous homogeneous reactors in the earliest days of the Metallurgical Lab. What was new was the idea that molten salt breeder reactors (MSBRs) could provide a primary source of energy at prices competitive with conventional power plants—and could do so safely and on a nationwide scale. To say the least, this was not a widely shared view, particularly among the mandarins of the AEC.

Not by nature a combative man, Weinberg had to use all his skills at persuasion and diplomacy to get the MSR off the ground. A realist, he understood that he was fighting not only for a specific R&D program but for the survival of Oak Ridge—his lab, his employees, and his colleagues and friends. By this time development of the light-water reactor had been turned over to Argonne National Lab, whose director, a close Fermi associate named Walter Zinn, was a fiercely territorial and stubborn scientist whose clashes with Rickover became legendary (and resulted, according to Weinberg, in Zinn's throwing Rickover out of Argonne, which is why the nuclear submarine engine was ultimately developed at the Westinghouse-run Bettis Laboratory).[22] For Oak Ridge to continue to receive funding and maintain its status as a premier research facility for nuclear physics and chemistry, Weinberg believed, the MSR had to be taken seriously.

"I argued that since industry was developing civilian nuclear power based on Rickover's light-water reactor technology, the national laboratories should pursue the long-term possibilities"—including the fast breeder under development at Argonne and the thorium–molten salt reactor at Oak Ridge. Kenneth Davis, director of the AEC's reactor development division, reluctantly concurred, and in 1956 Weinberg was given a relatively modest $2 million a year to assemble a team and perform the calculations for a theoretical MSR.

MacPherson, who had led the production of ultrapure graphite for the Hanford reactors, had been a student in the early days of the Oak Ridge School for Reactor Technology. One of the brilliant polymaths who seemed to be everywhere during the Manhattan Project era, he shared Weinberg's fascination with liquid thorium fuel. In 1959 MacPherson's team had proven the MSR concept on paper. Getting one built, though, seemed unlikely. Then MacPherson heard that the AEC had decided to fund as many as four "quick and dirty" reactor experiments, each costing less than $1 million. He wrote a proposal in a single night and submitted it to the AEC the next day. His quick and dirty proposal came in at $4 million, and inevitably the actual costs would eventually double. But the reactor division was persuaded.[23]

In MacPherson's design, the reactor vessel, containing an assembly of graphite rods 12.25 feet by 12.25 feet, would be made of a newly developed corrosion-resistant superalloy, developed for the nuclear aircraft program, that has since been trademarked as Hastelloy N. Through vertical channels in the graphite flowed a blend of uranium fluoride, thorium fluoride, and beryllium fluoride in a solution of 70 percent lithium fluoride. Heated to more than 1,000 degrees Fahrenheit, the fuel salt circulated from the core to four primary heat exchangers, with a barren (nonradioactive) coolant salt used as the intermediate fluid. The coolant salt, a mixture of lithium fluoride and beryllium fluoride, in turn dissipated heat through an air-cooled radiator. (In a power-producing MSR, the coolant salt would enter a secondary heat exchanger where water would be heated to generate steam to drive a turbine generator.)

Drawn up overnight in early 1959, this design differed only in minor detail from the designs for MSRs developed in the 1970s.[24] At this time there were two other competing forms of fluid fuel reactors: the aqueous homogeneous reactor, based on Wigner's ideas, which used a fuel solution of uranyl sulfate (UO_2SO_4) moderated by light water; and a reactor fueled by uranium metal dissolved in liquid bismuth, which grew out of concepts suggested by Leo Szilard. In 1959 the AEC realized that it had a multiplicity of proposed reactor designs on its hands and decided to narrow the field: a committee of outside experts was hired to evaluate and compare the three designs. MSRs won. "The Molten Salt Reactor has the highest probability of achieving technical feasibility," the experts concluded.[25]

It was at this point, in researching this history, that I stopped and said, "Let me get this straight." In 1959 the United States had a design for an MSR using thorium as its primary fuel. The nuclear power industry was in its infancy: the first commercial nuclear plant, at Shippingport, had gone on line only a year before. Molten salt, thorium-powered reactors had proven advantages over light-water uranium machines. Uranium was thought at the time to be scarce; everyone knew there was plenty of thorium. But uranium won out. It was like building an Edsel when you have the blueprint for a Porsche. A cavalcade of unforeseen and mostly unfortunate consequences has followed. It was one of the great technological missteps in history.

Named the director of Oak Ridge at age 40, Weinberg had the AEC's approval for a full-fledged MSR Experiment. He believed he had a clear path to achieving what he called "the holy grail" of nuclear power: a thorium-based thermal breeder reactor that could compete economically with existing forms of fossil fuel power generation. That, as he would say, was not to be.

TH90 · TH90 · TH90

BY THIS TIME HYMAN RICKOVER, having successfully launched the *Nautilus* in 1954, had embarked on the aging-dictator phase of his career. It's hard to overstate the influence that Rickover had gained over the AEC and nuclear reactor development by the beginning of the 1960s. He had turned the Naval Reactors Branch (NRB) into his personal fiefdom. In less than a decade he had built and launched ten nuclear subs, carrying the nuclear showdown to the most remote waters of the world. The technical handbooks and reports issued by the NRB became the standard works in the field, and its training programs turned out a couple of generations of nuclear power engineers—men who would go on to run the industry during its heyday of the 1960s and 1970s. The original nuclearati almost all trained at Rickover's feet. Single-handedly he had established the foundation for the nation's civilian and military reactor development.[26]

His power unchallenged, Rickover set about building a nuclear power industry in his own image. And that was the problem. Rickover's authoritarian style of leadership, his intolerance of dissent, and his valuing of efficiency over creativity and open discussion all bled into the roots of the nuclear power establishment. In Rickover's NRB, you didn't advance by thinking deeply about the problems of future energy supply or by delving into innovative reactor designs. You advanced by building nuclear submarines that didn't sink and by carrying out Rickover's designs. Rickover not only undermined and eliminated potential rivals and successors within the NRB; he did the same to potential competing nuclear programs. In some ways the field of nuclear engineering is only now recovering from Rickover's single-minded view of the technology.

Rickover's authority extended to the AEC, whose latest chair, the financier Lewis Strauss, was a Rickover enthusiast and free-market capitalist who railed against the "federal monopoly" on nuclear power. Believing that the market, and not federal regulators, should choose the best technology for commercial nuclear power, in 1956 Strauss helped engineer the defeat of the Gore-Holifield Amendment, which would have ordered the AEC to spend $400 million on six prototype reactors spread geographically across the country.[27] To Rickover and his suppliers, GE and Westinghouse, this would have been anathema.

The defeat of Gore-Holifield effectively "removed the government from research on new atomic power technology, which would now be left to industry."[28] *Industry*, in practice, meant the emerging leaders, Westinghouse and GE, which were already committed to light-water reactors designed for the Navy's submarine program.

TH90 • TH90 • TH90

THE COURSE OF NUCLEAR POWER HAD BEEN SET, but Weinberg and his team plowed ahead with the MSR Experiment. It went as well as it possibly could. Construction started in 1962 and the reactor went live in 1965, achieving its maximum power output of 7,500 kilowatts in March of the following year. As it turned out, to keep the construction simple and to save money, no thorium fluoride was included in the experimental reactor—a paradoxical decision since the entire rationale behind the program was its breeding potential, and under the circumstances a tactical error. But the experiment admirably demonstrated all the chief advantages of a thorium-based, liquid fuel reactor.

The fluoride salts remained stable, even at the very high temperatures inside the core. The machine was docile and easy to maintain. The circulating fuel could be reprocessed, and cleaned of undesirable fission products like xenon, during operation. The combination of the molten salt and the Hastelloy vessel proved, as expected, resistant to corrosion—although other problems with the integrity of materials cropped up, including a kind of "radiation hardening" that underlined the desirability of a core-and-blanket design that would prevent most fast neutrons from reaching the outer wall of the reactor vessel. A variety of successful experiments were carried out with the reactor, including studies of the behavior of tritium and the addition of plutonium to the system, showing that MSRs can consume a wide range of fuels.

In 1969 the MSR became the first reactor to run on uranium-233. The original radioactive material, U-235, was extracted from the fuel in a small chemical plant next to the reactor, where the uranium was boiled off with hydrogen fluoride. That process reduced the radioactivity of the fuel a billionfold, allowing the reprocessed uranium to be stored in unshielded steel containers. U-233, produced in a separate facility, was then circulated through the core. The codiscoverers of U-233, Ray Stoughton and Glenn Seaborg (by then chair of the AEC), were present when the reactor went

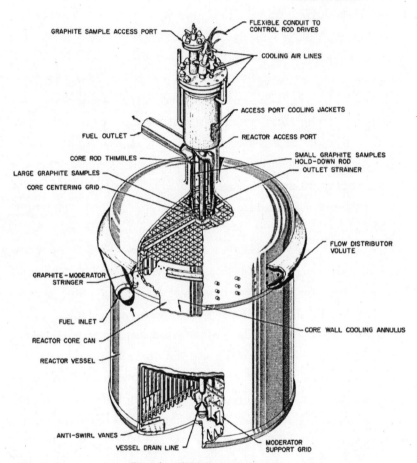

UNCLASSIFIED
ORNL-LR-DWG 61097R1A

GRAPHITE SAMPLE ACCESS PORT

FLEXIBLE CONDUIT TO
CONTROL ROD DRIVES

COOLING AIR LINES

ACCESS PORT COOLING JACKETS

FUEL OUTLET

REACTOR ACCESS PORT

CORE ROD THIMBLES

SMALL GRAPHITE SAMPLES
HOLD-DOWN ROD

LARGE GRAPHITE SAMPLES

OUTLET STRAINER

CORE CENTERING GRID

FLOW DISTRIBUTOR
VOLUTE

GRAPHITE-MODERATOR
STRINGER

FUEL INLET

REACTOR CORE CAN

CORE WALL COOLING ANNULUS

REACTOR VESSEL

ANTI-SWIRL VANES

VESSEL DRAIN LINE

MODERATOR
SUPPORT GRID

Fig. 6.   MSRE Reactor Vessel.

This diagram from the Oak Ridge National Laboratory archives shows the original molten salt reactor experiment, with a molten-salt core using graphite as the moderator. (Thorium Energy Alliance) critical with the thorium-derived fuel at its core. "It was a remarkable feat!" Weinberg exulted. "After such a success our confidence soared."[29]

Oak Ridge was now ready to build a much larger MSR as a commercial breeder prototype, including a blanket of thorium to produce the U-233 fuel. Designs for a large MSR demonstration plant were already underway. Weinberg expected the AEC's enthusiastic approval.

SIX

THE END OF NUCLEAR POWER

I n February 2011, I went to Oak Ridge to view Alvin Weinberg's papers. It had been a wet winter, and rainstorms lashed the hills surrounding the town. Oak Ridge today has become something of a bedroom community for Knoxville, 24 miles east on I–40; new condos litter the ridgelines, and the town itself has some trappings of a twenty-first-century American town: a health club, a Wal-Mart, a few chain restaurants strung along the Oak Ridge Turnpike. Weinberg would still recognize the place, but it would not be easy for a newcomer to perceive that one of the country's most sophisticated physics and chemistry labs sits just outside the city limits.

Some of Weinberg's papers now reside at the Howard Baker Center at the University of Tennessee in Knoxville. But the bulk is housed in a storage room the size of a walk-in closet at the Oak Ridge Children's Museum, a small schoolhouse-sized facility on West Outer Drive, up the hill from the center of town. No one could really explain to me how they ended up there; I'd been directed to the museum by Weinberg's son Richard, a retired biochemist. The staff welcomed me warmly, but they were slightly bemused. It was obvious that no one had asked to see Weinberg's archive in years. Oak Ridge National Laboratory honors its past, but it has not done a great job of preserving the records of its longest-serving director. This neglect, I thought, epitomized the neglected legacy of the man. Weinberg is a major figure in the history of nuclear power in this country, but, unlike Hyman Rickover or

Edward Teller or even David Lilienthal, Weinberg's name is virtually forgotten today.

The papers are stored in metal filing cabinets and cardboard boxes, piled literally to the ceiling of the windowless room. There's an index of sorts, and the cabinets are numbered, with subject-heading tabs on the file folders inside. Most papers are duplicates of Weinberg's correspondence, typed out on thin translucent ORNL letterhead paper, along with articles from scientific journals and technical reports from various laboratory programs, including the molten salt reactor (MSR). I spent an afternoon there, reading letters to editors, policy makers, Atomic Energy Commission (AEC) officials, and other scientists, many of them more than four decades old. I was hoping for a revelation, I'll admit, a smoking gun that would reveal some treachery at the birth of the nuclear power industry. I didn't find it exactly; most of the archive is routine administrative stuff. But buried in those cabinets is the inside story of a fierce bureaucratic struggle about the future of the MSR Experiment—and in a wider sense about the future of nuclear power. It was melancholy reading.

"Finally, I should like to take this opportunity to make known to you once again my objections to the drastic reduction in fluid fuels [funding] planned for fiscal year 1960," Weinberg wrote to Willard Libby, the AEC commissioner, in January 1959. " . . . To me it seems imprudent to cut fluid fuels in about half . . . instead of cutting back a little on several other much larger enterprises."[1]

That was almost a yearly lament. Weinberg spent most of the late 1950s and the 1960s pleading, arguing, and cajoling for more support for the fluid-fuels program and, once the aqueous homogeneous reactor project was canceled, specifically for MSRs. In 1956 ORNL's budget peaked at $60 million, with more than 4,300 people on staff. "We are the largest nuclear energy laboratory in the United States," Weinberg boasted, "and we are among the half-dozen largest technical institutions in the world."[2]

His exultation proved premature. The cancelation of the ill-conceived aircraft reactor in September 1957 resulted in a 20 percent across-the-board budget cut for the ORNL. Staffing was cut by 10 percent. The Eisenhower administration froze the laboratory's budget in 1957, forcing Weinberg to put off a major expansion that would have added a half-million square feet of new work space. These "cataclysmic setbacks" emphasized the new reality:

in the militarized, scientific culture of the Cold War, Oak Ridge was fighting for its survival.[3]

Weinberg believed that thorium-based fluid-fuel reactors would ensure the lab's central place in the emerging nuclear power industry. He was wrong. Officially the MSR Experiment lasted from 1959 to 1973, when it was canceled, only to be reinstated for reasons lost in the obscurity of long-ago energy policy in 1974 and then finally terminated for good in 1976. The most remarkable thing about the program is that, by all technical and economic measures, it was a resounding success. Politically, though, it was a dud. And year by year, letter by letter, Alvin Weinberg waged a lonely, ultimately losing battle to keep it alive.

Weinberg would say later that his lukewarm support for nuclear aircraft helped doom his quest to keep the MSR alive. In 1959 Weinberg was appointed to the White House Scientific Advisory Committee; his words had come to carry a weight they lacked when he was merely the director of Oak Ridge, however prestigious that institution was. In conversations that year with Herbert York, the former director of Livermore National Lab who had become the director of defense research and engineering at the Pentagon, Weinberg shared his skepticism. A year later John Kennedy shut down the Aircraft Nuclear Propulsion (ANP) program. "I don't have the slightest idea whether my conversation with Herb York had anything to do with the cancellation," Weinberg later recalled. "But Don Keirn, AEC's manager of ANP, and Kenneth Davis, who was the director of reactor development for AEC, must have thought so."[4]

After the cancelation, Keirn and Davis cornered Weinberg at the Roger Smith Hotel in Washington and accused him of treachery and disloyalty. "They chewed me out for about an hour," Weinberg recalled.[5] Permanently labeled an outsider and a malcontent, he would struggle with the AEC for the rest of his career at Oak Ridge. As laid out in his official correspondence, those struggles seem quixotic now. Weinberg was trying to change an entrenched militaristic culture with science and reason. Such battles seldom lead to victory. Nevertheless, he took up the search for sustainable nuclear power, based on molten salt thorium breeders, with a purity of purpose that no one else in the industry or the scientific community could match. In 1958 he had published an essay entitled "Power Breeding as a National Objective."

In it he argued that "current economics alone should not be the sole basis for choosing which reactor system to pursue. Efficient use of the raw materials of nuclear energy—uranium and thorium—was equally important."[6] Eventually they would become more important. Using the criterion of efficiency, there was no question which raw material made more sense. It was thorium, burned in molten salt reactors.

Weinberg's fervor for liquid-fuel reactors, and for the boundless potential for nuclear power in general, would lead him into some flawed projections. Indeed, he was ridiculed by early environmentalists. His memoirs take on an elegiac, even apologetic, tone as he looks back on the "nuclear euphoria" of the 1960s. He never stopped believing, though, that thorium-based nuclear power in fluid-fuel reactors could, in the long run, slake humanity's energy thirst for the foreseeable future.

TH90 • TH90 • TH90

IN 1958, AS PART OF HIS LONG twilight struggle to keep the fluid-fuel reactor program alive, Weinberg invited Eugene Wigner to visit Oak Ridge to review the state of the science. The second aqueous homogeneous reactor experiment had reached apparent failure when a hole formed in the interior wall of the core, made of a corrosion-resistant zirconium alloy. Jury-rigging a kind of periscope-and-mirror contraption, the engineers discovered that a region of the tank wall had melted when uranium settled out of the fuel solution and collected in that one spot. Stability of the fluid fuel had, from the earliest days of the concept, been a known hazard. Now it seemed that fuel instability might halt the fluid-fuel program in its tracks.

Wigner had returned to his post at Princeton; inviting him to Oak Ridge was a political risk on Weinberg's part. The former Oak Ridge research director was closely associated not only with his protégé Weinberg but with the fluid-fuel thorium breeder concept he'd originated before the war's end. Wigner's return was inevitably seen by some in Washington as an end-run around the AEC to lend to Weinberg's pet project the weight of Wigner's support. And, to some degree, they were right: Wigner, who spent December 22 and 23 in Oak Ridge touring the facility and being briefed on the fluid-fuel reactor program, came away convinced that the obstacles could be overcome. In the new year he wrote a long letter to Libby outlining his views. The letter refers specifically to aqueous homogeneous reactors, but its conclusions are applicable to all thorium-based fluid-fuel reactors. The letter is worth reviewing in detail for its candid look at the advantages and the challenges of thorium-fueled molten salt reactors. Much of Wigner's analysis remains relevant to today's LFTRs.

Wigner examined first the technical hurdles to fluid-fuel breeders and then the economics. Under "Technical Problems" he listed five elements of the technology: breeding, the blanket, the core, containment, and personnel. The last, Wigner argued, was a question of identifying and assigning the right talent. In the postwar years Oak Ridge had lost many of its brightest minds to academia and, in some cases, to industry. "It is my impression that the problems need the interest of some of the deeper thinkers in the laboratory."[7]

This was not an unprecedented observation from Wigner. A decade earlier, looking at the early postwar attempts to fashion a nuclear power reactor, he commented that reactor development had suffered from a lack of attention from "first-rate scientists." Now he was calling on the AEC to devote its best scientists to the development of new and advanced reactor designs. Since most of the Oak Ridge scientists had trained under Wigner and Weinberg, this was an understandable conclusion.

The design of the core received the most attention from Wigner. Fueled by a solution containing ten grams of enriched uranium per kilogram of heavy water, which circulated through its core at the rate of 400 gallons (1,450 liters) per minute, the reactor's fuel loop was comprised of the central core, a pressurizer, a separator, a steam generator, a circulating pump, and much interconnected piping. The core vessel was about a meter in diameter, centered inside a spherical pressure vessel made of stainless steel. A reflector blanket of heavy water—which in future molten salt designs would contain a thorium fluoride solution to breed U-233—filled the space between the two vessels. It was, according to the official history of ORNL, "perhaps the most exotic nuclear reactor ever built."

The challenge, as Wigner saw it, was to design a core in which the fluid (whether molten salt or aqueous) could achieve a high-enough temperature without losing stability—that is, without separating into two phases, one of them a uranium concentrate capable of melting the core wall. Unfortunately, this uranium-rich phase tended to separate and form a solid layer on the inner wall of the vessel. There were two ways to avoid phase separation: a design that involved "violent mixing and turbulence" in the core and a streamlined version that limited turbulence as the chain reaction occurred and the fluid flowed with a slow but more or less uniform velocity through the core. Wigner intuitively favored that streamlined system but acknowledged that "the weight of evidence at present" seemed to argue for high turbulence. In a broader sense, he said, the experimental design was simply not in accordance with the requirements of a thermal breeder. (In particular, the phase separation seemed to happen at screens that had been inserted into the core to help control the flow, when a layer of "rich phase" material adhered to the screens).[8]

"I am fully convinced that the problems of the core will be solved," Wigner wrote. "The reason that they have not yet been solved is principally due to two circumstances." First, the designers did not understand the phase diagram of the fluid in the core; second, "they followed their first impulses in design too strongly."

Wigner believed that adjusting the flow of fluid through the core would provide the velocity needed to prevent the uranium from settling on the walls, and he proposed removing the screens. He also proposed switching the inlet and outlet valves to reverse the fluid flow through the reactor. These measures would prove to be successful. And Wigner, who would win the Nobel Prize five years later, declared his unconditional support for Weinberg's efforts.

"It is my opinion that abandoning the program would be a monumental mistake," Wigner warned Libby.[9]

Reluctantly, Libby and the other AEC commissioners concurred. After several extended test runs using the new configuration, the reactor operated continuously in 1959 for 105 days—at the time a record for uninterrupted operation of a nuclear reactor. Wigner's intervention almost certainly saved fluid-fuel reactors—or, at least, postponed the day of their demise. The experiment conclusively demonstrated that such machines could generate power using far simpler systems than pressurized light-water reactors, with new fuel added and fission products removed continuously while the reactor continued to operate. Despite these accomplishments, near the end of that year another hole appeared on the core wall, and Westinghouse abandoned its plan to build a homogeneous reactor as a central power station for the Pennsylvania Power and Light Company. The AEC, as I've discussed, decided to place its fluid-fuel bets on the MSR, which would breed U-233 from thorium. As the 1960s dawned, Weinberg was confident this would happen.

TH90 • TH90 • TH90

NOT THAT HE WAS BLIND TO THE COMPETITION. "The boiler [i.e., pressurized-water system] bandwagon has developed so much pressure that everyone has climbed on it, pell mell," Weinberg complained in a letter to Wigner.[10]

By this time, 14 years into its existence, the AEC had become more of an arm of the Pentagon, and a distributor of funding to the competing national labs' reactor programs, than a true promoter and binding force for the development of peaceful atomic power. To be sure, the economic arguments for nuclear power from the dominant light-water reactor designs were weak: Commonwealth Edison had calculated in the mid-1950s that coal-fired plants would cost about $77 per kilowatt to build. Nuclear plants would cost $277. Following the "Atoms for Peace" speech, the Eisenhower administration pursued a two-pronged strategy whose inner contradictions would haunt U.S. foreign policy for decades: enabling the spread of nuclear technology, mostly through light-water reactors, while continuing a massive buildup of the nuclear weapons stockpile, including the "Super," the hydrogen bomb first tested during the Truman administration in 1952. Light-water reactors, by their nature, required the enrichment of natural uranium. Pursued to its logical end, the enrichment of uranium was the first, and in many ways the most difficult, step toward obtaining nuclear weapons. Among the countries that benefited from U.S. atomic technology proliferation were India and, eventually, China.

The development of a strong nuclear industry, said a National Security Council memo, "is a prerequisite to maintaining [America's] lead in the atomic field."[11]

Established by the scientists of the Manhattan Project, that lead proved to be short lived.

By the end of the decade Rickover's light-water reactors had suffered no serious accidents. The nuclear sub program had done more than anything else to demonstrate the ability of nuclear plants to provide safe and continuous power under demanding conditions. Many scientists, though, remained concerned about the safety of nuclear plants, and the congressional Joint Committee on Atomic Energy had commissioned a probability study (the first of many) to establish the level of risk. Published in early 1957, the report was sobering: While the chances of a serious accident were considered remote, a worst-case scenario could result in 3,400 deaths, 43,000 injuries, and $7 billion in property damage.[12] Like all such studies, it was shelved after generating a few scary headlines. And, officially, support for nuclear power continued unabated.

Unfortunately, this political enthusiasm failed at first to translate into commercial activity. By the end of the 1950s the United States, the nation that had harnessed nuclear fission and invented the first nuclear reactor, the adopted home of Fermi and Wigner and Teller and Szilard, was actually behind other nations in the development of power-generating reactors. In 1954, with Washington gripped by the Army-McCarthy hearings, the first civilian nuclear power plant, a five-megawatt plant south of Moscow, was started up by the Soviets. The British, whom the United States had essentially tossed out of the nuclear effort after Hiroshima, had built a 100-megawatt plant at Calder Hall. Lewis Strauss, the AEC chair enthralled by Rickover, had predicted that future Americans would enjoy nuclear power "too cheap to meter." But, hindered by an AEC primarily focused on the military applications of nuclear energy and by an industry still dependent on the Pentagon for both staffing and financial backing, nuclear power in the United States at the time of John F. Kennedy's inauguration had failed to fulfill the promise foreseen by its early enthusiasts.

That situation would change, the AEC was convinced, with the emergence of breeder reactors. When the Soviets, at the first United Nations conference on Peaceful Uses of Atomic Energy in Geneva in 1955, announced plans for a breeder reactor that would generate power, Strauss immediately recruited the Great Lakes utility Detroit Edison to apply for an AEC license to build a commercial breeder. That project, on the shore of Lake Erie, would become the ill-fated Fermi 1 liquid metal breeder reactor. And the misguided efforts to build a safe, efficient breeder reactor would ultimately doom Weinberg's thorium-based MSR.

TH90 • TH90 • TH90

WHEN I FIRST STARTED RESEARCHING the history of thorium reactors, I received—and mostly swallowed—a straightforward tale: The United States abandoned thorium reactors because they didn't produce plutonium for bombs. Full stop. Populated by pacifist scientists on the one hand and Machiavellian warmongers on the other, this made an appealingly symmetrical story. But it wasn't the whole truth. In fact, I realized as I photocopied old letters that the truth, as it tends to be, was much more complicated. In the early 1950s molten salt thorium-fueled reactors lost out to light-water reactors because the latter were more developed and Rickover wanted to build nuclear subs as fast as the Navy would fund them. Later, in the 1960s, MSRs lost out to liquid-metal breeders for a constellation of reasons that included, but were not limited to, the Cold War demand for nuclear warheads. This was a harder weave to unravel and a harder story to tell.

Despite lukewarm support from the AEC, the Molten Salt Reactor Experiment (MSRE) proceeded in the new decade as Weinberg's men gradually solved the materials and design issues that had plagued the aqueous homogeneous technology. Molten salts were more stable than aqueous solutions; they provided a clear path to U-233 breeding from thorium; and they had such a high boiling point (680 degrees Fahrenheit) that the reactor could operate at atmospheric pressure—a fundamental advantage over not only pressurized water reactors but also the aqueous homogeneous system. The "two-fluid" design of the MSR involved complicated plumbing to keep the fuel salt and the blanket salt separate but interlaced within the graphite core, but it remained a simpler machine than the pressurized water reactors (PWRs) of the day. First brought to criticality on June 1, 1965, the MSR experiment achieved a number of important milestones: it ran continuously for six months, it demonstrated the practicality of molten salts in a nuclear reactor, and, on October 2, 1968, it went critical with uranium-233, bred from thorium. Four months later it reached full power (8 megawatts) using U-233. To Weinberg, Mac MacPherson, and their team, it seemed an ideal system, the Platonic version of a nuclear reactor.

"Here we had a high-temperature fluid-fuel reactor that operated reliably and, even in the primitive embodiment represented by MSRE, had remarkably low fuel costs," Weinberg wrote.[13]

In December 1969 the MSR was shut down to make way for what Weinberg believed would be more advanced molten salt designs and a full-fledged demonstration plant.

To be sure, there were technical difficulties. Various mechanical problems delayed start-up for a period in 1964–65. During the experiment three persistent challenges revealed themselves. One, the Hastelloy-N walls of the reactor vessel, subjected to constant neutron bombardment for a period of years, suffered "radiation hardening" related to the buildup of helium atoms. The materials scientists at Oak Ridge later developed new alloys containing fine carbide precipitates that would hold the helium and limit the hardening. Future designs would include a blanket of fertile thorium, which would help protect the vessel wall.

The second problem also had to do with Hastelloy N. The pipes for the MSR developed tiny cracks on their interior surfaces caused by the fission product tellurium. Adjusting the chemistry of the fuel salt (increasing the percentage of UF_3, as opposed to UF_4) eliminated the cracking.

Third was the buildup of tritium, also a product of the fission of uranium-235 and uranium-233. A radioactive isotope of hydrogen, tritium penetrates metals easily and could be vented to the atmosphere through the steam generators. MacPherson's men spent months on the tritium problem. Ultimately they realized that the intermediate salt coolant would capture the tritium, which could then be removed during the reprocessing system that was in place to purify the system of other poisons, such as xenon.

All those problems proved to be solvable. The extended full-power run of the MSR experiment—87 percent of the time during 15 months of operation—was a feat of nuclear engineering in itself. "When measured against the yardstick of other reactors in a comparable stage of development, it is seen to be indeed remarkable," wrote the Oak Ridge scientists Paul Haubenreich and Richard Engel in a 1970 account of the experiment in the journal *Nuclear Applications & Technology*.[14]

Nevertheless, the demo plant was never built. The reasons why remain cloaked in myth, speculation, and institutional amnesia. At the center of the controversial decision, though, was Milton Shaw.

If there's a Machiavelli in this story, it is Shaw. A native of Knoxville, he studied mechanical engineering at the University of Tennessee before joining the Navy, too late for World War II. After postgraduate study at the Navy Propulsion School at Cornell, he was assigned to the Pacific theater, where he served as an engineering officer until the surrender of Japan. At some point he sought out Rickover, who recognized in Shaw a perfect subordinate: bright, almost as driven as his superior, and equally pugnacious. Rickover sent him to the School of Reactor Technology at Oak Ridge, where Shaw was an average student. For 11 years, through 1961, he reported directly to Rickover in the Naval Reactors Bureau, becoming known as the admiral's chief henchman, Beria to Rickover's Stalin. Shaw was the project leader for both the *Nautilus* and the *Enterprise,* the first nuclear aircraft carrier, launched in 1960.

A Rickover man to his core, Shaw inherited Rickover's abrasive, autocratic leadership style and his intellectual inflexibility. In 1961 Shaw became the senior assistant to the Navy assistant secretary in charge of all research and development for the Navy and the Marines, and in 1964 he joined the AEC as director of reactor R&D. In that capacity he accomplished the most significant feat of his career, aside from naval nuclear propulsion: killing the molten salt reactor and eliminating all competition to the liquid metal fast breeder reactor as the next stage of advanced nuclear power development.

Shaw always viewed Oak Ridge with distaste, regarding it as a lab full of prima donna propeller-heads and possible pinkos. He was especially scornful of Alvin Weinberg. Shaw "was accused of using his authority to destroy the lab," wrote Charles Barton Jr., son of one of the preeminent nuclear chemists at Oak Ridge. "I do not know if that was his intention, but he certainly succeeded in destroying the Reactor Chemistry Division."[15]

Weinberg, naturally, was more diplomatic. Late in life, in an interview with two journalists associated with ORNL, Weinberg described the man who effectively ended his Oak Ridge career: "Milton Shaw had a singleness of purpose. In many ways I admired him, and in many ways he drove me nutty. He had a single-minded commitment to do what he was told to do, which was to get the Clinch River Breeder Reactor built."[16]

Weinberg, who was never convinced that liquid metal fast breeders could operate safely (and who was proved correct), disagreed with the Clinch River plan. "I think the [Atomic Energy] Commission decided that my views were out of touch with the way the nuclear industry was actually going."[17]

TH90 • TH90 • TH90

THE 1960S WERE WEINBERGER'S HEYDAY, his period of irrational exuberance. In Europe, unlike his native country, he was feted as a hero and a pioneer. He got an early taste of the Continent's enthusiasm for advanced nuclear technology when he was invited to speak at the national laboratories of France, Great Britain, Germany, Denmark, Sweden, Italy, and just about every other country in Western Europe. The countries that had given birth to the incomparable scientific minds who ignited the atomic revolution—minus Hungary, which was frozen in the shadow of the Iron Curtain—now wanted to taste the revolution's freedoms. Eagerly feeding what he called "international euphoria," Weinberg became the new industry's leading scientific evangelist. "I found myself becoming a sort of unofficial nuclear ambassador, invited by many of the fledgling European nuclear groups to visit and tell them about reactor developments at Oak Ridge." Scientists in Asia, particularly Japan—the only country ever to be the target of a nuclear attack, and a place where the power of nuclear fission is given unique respect—likewise wanted to hear about the U.S. program. Even the Kurchatov Nuclear Institute outside Moscow, named for the father of the Soviet nuclear weapons program, welcomed the nuclear ambassador. Igor Kurchatov himself ("the Soviet equivalent of Robert Oppenheimer, Ernest Lawrence, and Enrico Fermi all in one") introduced Weinberg to an audience of eager note-taking communist nuclear engineers.[18]

Everywhere he went, Weinberg preached the gospel of thorium. "I extolled the promise if not the virtue of nuclear power, especially if it was based on the thorium-burning molten-salt breeder."[19]

In Moscow, as in Tokyo and Berlin and London, Weinberg found kindred listeners. Europe had no oil, only expensive coal mines; its power was costly, and its fledgling nuclear programs were firmly under civilian control. Weinberg's travels gave him an outsider's perspective on the system in which he worked: a military-industrial complex driven primarily by bombs and nuclear propulsion. "Even the naval program, though carried out adequately by GE, Westinghouse, and Argonne, was very much in the grip of Rickover and his people." The best scientists had retreated to the universities or found work designing ever-more infernal machines of destruction. Nuclear power got, if not the dregs, a large share of the second-raters. In Europe it was different. Visiting nuclear stations in the Old World, Weinberg remarked, "I am struck by the technical sophistication of the European nuclear-plant managers as compared to the American plant managers I have met."[20]

Encouraged by this seriousness and sophistication, Weinberg proceeded to make speeches that would sound overly optimistic, even absurd, two decades later. The future of humanity, he claimed, lay in "burning the rocks and burning the sea." Building reactors that used the virtually limitless supply of thorium in Earth's crust, he said, would be like burning the rocks. In future centuries, he predicted, we would "burn the sea" in fusion reactors run on the deuterium from seawater—controlled versions of the thermonuclear bomb.

Fed by the dream of inexhaustible, inexpensive energy, Weinberg's projections became grandiose. The Oak Ridge scientists studied the "construction of giant agro-industrial complexes built around nuclear reactors . . . as a means of providing food and jobs for millions of persons in underdeveloped countries," the *New York Times* reported in 1968. A complex built around thorium breeders could sustain 100,000 farmers and laborers, "feed five million others and export fertilizers to grow food for 50 million additional people." A 2,000 megawatt nuclear plant would desalt a billion gallons of water a day, irrigating vast plantations in the desert. Nuclear cities, enthused Weinberg, "could be the Apollo project of the nineteen-seventies."[21]

The response of men like Rickover and Shaw to such fantasies is not hard to imagine. President John Kennedy, though, was swept away by Weinberg's imagined nucleoparadise. In a speech just weeks before his assassination, JFK foresaw the day when vast nuclear stations would generate abundant electricity as well as a bottomless supply of freshwater: "We will have before this decade is out or sooner a tremendous nuclear reactor which makes electricity and at the same time gets fresh water from salt water at a competitive price."[22]

Weinberg even believed that thorium breeders would unravel the knot of Middle Eastern politics. With the encouragement of Howard Baker, the Republican senator from Tennessee, Oak Ridge produced a multivolume plan for nuclear agroindustrial complexes to be built at Al-Hammam, near Alexandria; in Israel's western Negev desert; and an "international project" in the Gaza Strip. Technology would unite Arab and Israeli in a nuclear utopia.

Israeli ambassador Yitzhak Rabin heard Weinberg out during a stopover at the Knoxville airport and, with Levantine skepticism, congratulated him for his chutzpah.

Others were also incredulous. By the late 1960s the publication of Peter Matthieson's *Wildlife in America* (1959) and Rachel Carson's *Silent Spring* (1962) had given rise to an environmental movement that increasingly decried nuclear power as an unacceptably risky energy source. Even some nuclear scientists, such as MIT's Henry Kendall, a future Nobel laureate in physics and the founder of the Union of Concerned Scientists, had begun to turn against the atom as a source of electricity. After a 1970 speech in Vienna marking the twenty-fifth anniversary of the International Atomic Energy Agency—an occasion when Weinberg extolled the potential for limitless nuclear power and once again declared that "a visionary world of abundance" lay before us—he heard muttering in the corridors. Some audience members, many of whom were actually building nuclear plants and finding them far more expensive than the best-case cost scenarios laid out by Weinberg, called him a charlatan and accused him of indulging in "gross over-optimism."[23] The tide was turning against him, but Weinberg, naively convinced that the thorium breeder would sweep away all possible objections, had not sensed it.

TH90 • TH90 • TH90

DESPITE THE SUCCESS OF THE MSR EXPERIMENT, Milton Shaw issued a series of critical assessments of the technology culminating in a 1972 report designated WASH–1222. Titled "An Evaluation of the Molten Salt Breeder Reactor," this document has taken on, in the thorium movement, a significance akin to the contemporaneous Pentagon Papers among antiwar activists during the Vietnam War. In a remarkable piece of bureaucratic jiujitsu, Shaw in WASH–1222 candidly acknowledges the thorium-based MSR's obvious benefits and then summarily concludes that it should be abandoned. In classical rhetorical terms, Shaw's argument is a syllogism: he takes two premises and from them infers a conclusion. But his logic is flawed. His first premise is that "it has not been proven that molten-salt reactors can work at commercial scales." His second premise states, "Bringing the molten-salt breeder to commercial production would require large amounts of time and money, and the solution of daunting technological problems."[24] The conclusion that the thorium breeder should be terminated does not follow. It was the first of many versions of what would become a familiar argument: *It hasn't been done before, and doing it would be challenging. So we shouldn't try it at all.*

"It is noted that this concept has several unique and desirable features; at the same time, it is characterized by both complex technological and practical engineering problems which are specific to fluid-fueled reactors and for which solutions have not been developed." Those problems were "different in kind and magnitude from those commonly associated with solid fuel breeder reactors."[25] Shaw estimated that $150 million had been spent to that point on the molten salt breeder reactor (MSBR) program, a figure that was almost certainly inflated. Total government investment to bring the molten salt breeder to commercial production would rise to about $2 billion, he said. (ORNL's own proposal for future development of the molten salt breeder reactor, submitted in 1972, called for about $3.8 billion over 11 years.)

Of course, at that point the AEC was already engaged in developing a reactor that presented "problems different in kind and in magnitude" from conventional light-water reactors and that would consume far more than $2 billion during the ensuing decade. That technology was the liquid metal fast breeder reactor (LMFBR), embodied in the Clinch River Breeder Reactor program, on which the United States would eventually spend $8 billion without a single spade turned for construction. All the evidence today indicates that Shaw seriously overestimated the maturity of competing technologies (indeed, problems with the LWR itself would cost operators tens of billions of dollars in the next decade), and he vastly overestimated the challenges for the molten salt breeder.

Nonetheless, the Nixon administration had been sold on liquid metal breeders. In 1971, in the first-ever special presidential message to Congress on energy, Richard Nixon declared that "our best hope today for meeting the nation's growing demand for economical clean energy lies with the fast breeder reactor." Nixon requested an additional $27 million for fiscal year 1972 for the LMFBR effort and committed to a full demonstration of the liquid metal fast breeder reactor by 1980.[26]

Shaw also notes that further development of the thorium MSR would require significant resources, particularly "qualified engineering and technical management personnel and proof-test facilities" that were in acute shortage at the time. This was nonsense; Shaw was deciding among competing technologies by stating the obvious fact that not all warranted the same level of government and industry commitment. He had already, since joining the AEC in 1964, worked tirelessly to whittle away the financial and personnel resources available to Weinberg's MSR program—indeed, to throttle the R&D capabilities of Oak Ridge as a whole. Shaw had settled the matter in his own mind; liquid metal breeders would go forward, and molten salt breeders would wither. WASH–1022 was a whitewash that provided flimsy justification for a conclusion that had been reached years before.

In arriving at that conclusion, Shaw dismissed the level of private-industry interest in the MSBR as negligible. This was demonstrably false. The 1970 *Minerals Yearbook,* published by the Bureau of Mines, reported that "there was . . . a significant increase in private efforts involving this concept. The Molten Salt Breeder Reactor Associates, an association of five electric utility companies and a consulting engineering firm, completed Phase I of their study of the MSBR. In addition, 15 utility companies and six major industrial companies formed the Molten Salt Group, which will jointly study MSBR technology, including the feasibility of thorium as a fuel."[27]

Shaw's reasoning was perfectly circular: *Private industry will not invest in the MSBR as a commercial venture without the support of the government. We, the government, won't support it. Thus private industry won't invest in it.*

Shaw himself had destroyed the resources that were required to bring the molten salt breeder program to fruition. In WASH–1222 he simply made explicit a policy he had pursued since joining the AEC. Molten salt reactors and thorium power never recovered from the blow. The program was officially terminated in 1973 in a letter to Weinberg that echoed Shaw's dubious logic: "Among the many outstanding achievements in the MSR research and development program, the highly successful performance of the Molten Salt Reactor Experiment will long endure as the most significant," wrote R. E. Hollingsworth, general manager of the Reactor Division of the AEC. "However, the additional commitments by the Government of the costs and other resources which would be required to demonstrate the potential of the concept for commercial applications would be very high."[28]

Congratulations: your program was an outstanding success. Now we're shutting it down.

TH90 • TH90 • TH90

I FOUND THE SURVIVING SCIENTISTS who worked on the MSRE at Oak Ridge sur-
prisingly fatalistic, today, about the fate of their program. "There was nothing
we could do about it," Richard Engel told me in an interview. "The matter
had been decided in Washington." It was an unfortunate decision, but they
had other work to do, and families to feed, and soon they were on to other
programs. Wishing otherwise was irrational.

Four decades on, though, the current state of our power supply and
Earth's climate demand understanding: Why was the molten salt breeder—a
technology with demonstrable benefits over competing technologies, which
had been proven to work, which had the support of industry and was certainly more economical than its competitor, the liquid metal fast breeder,
shut down? Shaw may have been rigid and blinkered, but he was no fool.
What could have been done differently?

Mac MacPherson, whose career was more tightly aligned with the MSR
program than even Weinberg's, believed that tactical errors made early on by
the Oak Ridge leadership led to the program's demise. First, Weinberg had
failed to drum up support for the technology at other labs and in Washington, D.C.: "Only in Oak Ridge . . . was the technology really understood
and appreciated." Outside ORNL, predictions for the ultimate potential of
the molten salt breeder were considered outlandish. In fact, Weinberg and
his men had made the mistake of proceeding too cautiously, MacPherson
argued: The scientists leading the molten salt program chose to wait until
the MSR experiment proved successful before expanding the program to a
full-fledged demonstration plant; by the time that happened, the liquid metal
breeder had already gained momentum, along with the support of Shaw and
his political mentor, Representative Chet Holifield, Democrat of California,
the powerful chair of the Joint Committee on Atomic Energy. In a politicized
arena, prudence was the enemy of innovation. The liquid metal bandwagon
was rolling along, and "it was asking too much of human nature to expect
them to believe that a much less expensive program could be effective in developing a competing system."[29]

Weinberg acknowledged the political shortcomings of the MSBR program, but he added another, more technical reason. Liquid metal breeders

were an extension of the extant, and successful, light-water reactors. Molten salt technology, though simpler, was new and strange. Milton Shaw, a man whose entire intellectual and managerial background was in the naval reactors effort, believed that LWRs were a proven, mature technology, that nuclear safety was a nonissue, and that liquid metal breeders were the logical—indeed, the only—next step. He was incapable of entertaining the idea that a completely different and, in his eyes, exotic technology—even one that had been conceived during the war at the dawn of the Atomic Age—could present surpassing technical and economic advantages over what he saw as a perfectly fine system that already enjoyed broad institutional support. His personal distaste for Weinberg, who emphasized the long-term costs and risks of nuclear power, and for the unruly boys at Oak Ridge, was a separate but intertwined issue.

In a 2010 article on molten salt reactors in *Mechanical Engineering* magazine, David LeBlanc—a physicist at Carleton University in Ontario and the founder of Ottawa Valley Research Associates, a start-up created to advance new MSR designs—mentions another, more sinister reason for the MSBR's cancellation. "Finally, and more speculatively, is the theory that the MSR was killed because it didn't produce plutonium, which was a military objective."[30]

The theory is now repudiated not only by more circumspect scientists like Jess Gehin, a senior program manager for reactor technology at Oak Ridge, but by many in the thorium movement. The evidence that Shaw and the AEC killed the MSBR specifically because it wasn't an efficient producer of weapons-grade plutonium is thin. But that misses the point.

Light-water reactors and their younger cousin, the liquid metal breeder, won out because of technological intransigence rooted in the military origins of the U.S. nuclear program. The men who controlled nuclear power in this country had come, almost uniformly, out of Rickover's nuclear Navy; they saw civilian nuclear power as, at most, an extension of the nuclear weapons program—a congenial side benefit, as it were, to an overwhelming strategic imperative. The progressive militarization of U.S. society and industry—cited most famously in President Eisenhower's valedictory speech of 1961 in which he gave a gloomy, prescient forecast of the rise of the military-industrial complex, warning the nation against placing its trust in generals, admirals, and their civilian contractors—prevented the nuclearati from seriously considering a technology that had little to do with nuclear weapons and that, moreover, had been advanced by a laboratory director with suspected antiwar and environmentalist leanings. Claiming that the cancelation of the program was a direct result of the demand for nuclear weapons is an oversimplification; saying that the two had nothing to do with each other is naive.

Ultimately, the AEC mandarins concluded that the molten salt breeder reactor couldn't be done because it had never been done. Coming so soon after the triumphant landing of an American manned spacecraft on the moon—an effort that at the time of John Kennedy's 1961 address to Congress was far less advanced than the molten salt reactor (and of far more doubtful benefit to the nation, and to humanity)—this was a dispiriting conclusion.

TH90 • TH90 • TH90

THE AFTERMATH WAS EVEN MORE SO. Each of the three main figures in this drama lost his job, against his will and in humiliating fashion.

Weinberg was the first to go. In fact, his position had been deteriorating for some time—not only because of his support for molten salt reactors but because of his focus on nuclear reactor safety. As far as back as the mid-1950s, George Parker of Oak Ridge was measuring the lingering radioactivity of irradiated, melted fuel elements. In the early 1960s Parker, under Weinberg's direction, instituted a series of annual international conferences on reactor safety issues. Shortly after Shaw came to power at the SEC, the conferences were terminated. The journal *Nuclear Safety*, published by ORNL, became the primary vehicle for new studies on the subject. And Alvin Weinberg became known as the conscience of the nuclear power establishment—a role that hardly endeared him to Milt Shaw or Chet Holifield, who by this time held virtually unchallenged power over the AEC.

In his most famous statement, Weinberg pithily expressed the dilemma in which nuclear power advocates found themselves: "We nuclear people have made a Faustian bargain with society. On the one hand, we offer . . . an inexhaustible source of energy. But the price that we demand of society for this magical energy source is both a vigilance and a longevity of our social institutions that we are quite unaccustomed to."[31]

Weinberg seems to have gotten the message late that he was on the losing side of the debate. In 1972, in a discussion with Holifield and ORNL deputy director Floyd Culler, Holifield blurted out, "Alvin, if you're so concerned about the safety of reactors, then I think it may be time for you to leave nuclear energy."[32]

The final straw was when Weinberg committed the sin of confessing his doubts to the opposition. Angered by Holifield's rebuke, Weinberg agreed to have dinner with Ralph Nader, whose sister Claire worked with Weinberg at Oak Ridge. Nader at that time was becoming one of the loudest voices against nuclear power, and Weinberg was frank with him. Weinberg insisted to Nader that the chances of a serious nuclear accident were so small as to be inconsequential compared with other forms of energy production—including fossil fuels, which produce carbon dioxide. But, he added, the AEC was not taking the issue seriously enough.

Weinberg would later regret his conversation with Nader. But the damage had been done. It fell to John Swartout, a former Weinberg deputy who was now the vice president for research at Union Carbide, which operated Oak Ridge under federal contract, to deliver the bad news. After 18 years as director of the world's leading laboratory for research on advanced nuclear technology, Weinberg was out. The nuclear era he had envisioned, worked for, and espoused was placed on indefinite hold as of 1973.

That was the year everything changed. "The most pivotal year in energy history," according to the U.S. Energy Information Administration.[33] It was the year the Arab oil sheikhdoms cut off supplies to the West, establishing the hegemony of the Organization of Petroleum-Exporting Countries (OPEC), and setting in motion the petroleum-fueled conflicts that still roil the world today. In 1973 the U.S. nuclear industry signed contracts for 41 new nuke plants—all were uranium-powered light-water reactors—the industry's highwater mark and, although it was not apparent at the time, its final flourishing. No reactor ordered after 1973 was ever brought into operation. It was also the year that funding for the Clinch River Breeder Reactor began, having been approved by Congress a year previously. It would be canceled by Jimmy Carter after the Three Mile Island accident, briefly revived by the Reagan administration, and shelved for good in 1983.

In October, Egypt and Syria invaded Israel, triggering the Yom Kippur War and leading, with a grim inexorability, to the OPEC oil embargo that launched the first energy crisis and awoke Americans to the tenuousness of their energy sources. The following month Nixon announced "Project Independence," calling for energy self-reliance by 1980—a goal centered on liquid metal breeder technology. It would come no closer to fulfillment than the Clinch River reactor project.

And it was the year that thorium-based molten salt reactors died, making thorium power one of the great what-ifs of the Atomic Age and taking with them Alvin Weinberg's vision of a green and pleasant land dotted by safe, clean reactors.

Shaw was the next to go. In 1973, responding to environmentalists' outcry, Nixon appointed Dixy Lee Ray, a marine biologist and former director of the Pacific Science Center, as chair of the AEC. Ray had little patience for Shaw's Stalinist tactics or for the Navy's old-boy network that dominated the industry, and she essentially removed nuclear safety from his purview. Furious, he resigned. Famous as the man who launched the *Nautilus,* he enjoyed a comfortable retirement as a consultant and a visiting professor at MIT and Carnegie-Mellon. Even as Three-Mile Island and Chernobyl poisoned public perceptions of nuclear power, he never publicly reconsidered his positions on reactor safety or on molten salt reactor technology.

Oddly his mentor did reconsider. In a 1984 interview with Diane Sawyer, President Jimmy Carter (himself a veteran of the Naval Reactors Branch) recalled a conversation he had with Admiral Rickover on one of the nuclear submarines he had helped create.

"I wish that nuclear power had never been discovered," Rickover told his former junior officer. When Carter protested, "Admiral, this is your life," Rickover replied, "I would forego all the accomplishments of my life, and I would be willing to forego all the advantages of nuclear power to propel ships, for medical research and for every other purpose of generating electric power, if we could have avoided the evolution of atomic explosives."[34]

For a man who never in his life second-guessed a decision he had made, this was a remarkable bit of self-reflection. In his public life, though, the bantam admiral had no time for such doubts. He resisted numerous attempts by Congress and successive administrations to oust him, and he clung to his naval career like a mariner clutching a tiller in a storm. He had seen the moment of his greatness flicker long before, but he refused to acknowledge the inevitable. When the end finally came, it was predictably unseemly. Accused by a Navy ethics board of accepting gifts from nuclear contractors, including General Dynamics and GE, Rickover clashed with Navy Secretary John Lehman, a distinguished former naval aviator who had come to view the "Rickover cult" as a major obstacle to renewing a service plagued by a loss of strategic vision, plummeting morale, and overly cozy relations with suppliers. In a now-famous Oval Office meeting with Lehman, President Ronald Reagan, and Defense Secretary Caspar Weinberger, Rickover called Lehman a "piss-ant" and a "Goddamn liar" and claimed to be "the only one in the government who keeps [the contractors] from robbing the taxpayers."[35]

All to no avail. Reagan informed the 83-year-old admiral that his career was over. Rickover spurned all ceremony and raised to the Navy that had been his life for more than 60 years the same finger that he'd given it since his days at Annapolis, two world wars ago. He died four years later and was buried in Arlington National Cemetery. It's safe to say that his like will not tread the decks of the nation's submarines again.

As for Alvin Weinberg, after a brief, stormy, and fruitless year in the capital as director of the newly created U.S. Office of Energy Research and Development—a federal appendage every bit as toothless as it sounds—he returned to distinguished exile in Oak Ridge to teach, write, and serve as the éminence grise of nuclear power for the rest of his life. His memoir contains a poignant recollection of his solitary late-night rambles through the streets of Washington during "the worst year of my life," haunting the city like the ghost of a more enlightened and more promising age. His post-ORNL years were satisfying in many ways, but his biography is clearly that of a thwarted man. Near the end of his life (he died in 2006, at age 91), he spoke and wrote often of "the second nuclear era," in which he never lost faith. He was a man of great foresight, moral force, and patience. His flaw was his idealism. He believed that science would rout ignorance, that reason would triumph over political might and personal ambition, and that his antagonists' motives were as pure as his own. A more willful and ruthless man—a Rickover, or a Shaw—might have been able to see thorium molten salt breeders through to their eventual triumph.

"During my life I have witnessed extraordinary feats of human ingenuity," Weinberg wrote. "I believe that this struggling ingenuity will be equal to the task of creating the Second Nuclear Era.

"My only regret is that I will not be here to witness its success."[36]

SEVEN

THE ASIAN NUCLEAR POWER RACE

The island city of Mumbai, India's cultural and business center, droops like an elephant's trunk into the Arabian Sea. The Trombay district in the northeast of the city occupies the elephant's forehead. Inhabited for centuries by fishermen who plied Thane Creek, which forms Mumbai's eastern boundary, Trombay was once an island itself, but successive reclamation projects have joined it to the main mass of the city. Nevertheless it's still a separate place, quieter and less chaotic than Mumbai proper. It's known for the Palipada *masjid*, or mosque, one of the oldest on the Indian subcontinent, and for several seventeenth-century Portuguese churches, built by the earliest European explorers, that now lie in ruins. Trombay has one of India's most diverse population mixes, Tamils and Maharashtrians living in relatively harmonious proximity to Punjabis, Sindhis, and Keralites.

It's also known for its warehouses and industrial complexes, and for the foremost scientific institution in India: the Bhabha Atomic Research Centre (BARC), the heart of India's ambitious thorium power program.

India is surely the only nation in the world to place a nuclear research center with at least seven operating reactors on the edge of one of its most populous cities. Mumbai has a population of almost 21 million, a great many of whom live within a dozen or so kilometers of the Bhabha Centre. It's as if Los Alamos were located in Brooklyn. There has never been a serious accident at Bhabha. If there ever were, it would make Chernobyl look like a brushfire.

Even the Soviets had the sense to locate their major reactor complexes far from urban centers.

Surrounded by circular gardens and artificial ponds, with towering smokestacks and a huge containment dome, BARC bears a strong resemblance in silhouette to a citadel of medieval Islam, dominated by domed mosques and towers for the muezzin who call the faithful to prayer. Bhabha, though, is a decidedly secular temple. The thorium-powered second nuclear era that Alvin Weinberg prophesied is being forged here. With a population of 1.21 billion that adds another 20 million or so souls each year, and an annual economic growth rate of about 8.5 percent, India is one of the most energy-hungry nations on Earth. Today India gets less than 3 percent of its electricity from nuclear plants, relying heavily on coal and crude oil imports to fuel its economic boom. The country has 27 nuclear reactors today, generating about 4.7 gigawatts. The government of Prime Minister Manhoman Singh has said it will build up to 62 reactors by 2025, generating about 63 gigawatts and raising nuclear energy's share of the country's electricity production to 25 percent. And most of those reactors will be advanced heavy-water converter reactors, running on an inexhaustible supply of thorium.

India, which has either the world's largest or second-largest reserves of thorium (in competition with Australia, depending on who's counting and how recently the calculation was made), is the only country in the world with a detailed, funded, government-approved plan to base its nuclear power industry on thorium-fueled reactors. If there's a best hope for thorium power, it's in India. But there are significant institutional, social, and technological barriers to the Indian program that raise serious doubts as to whether the country's grand thorium plans will ever be realized.

This is a matter of importance not just to Indians. The degree to which India and its neighbor to the northeast, China, can wean themselves from fossil fuels and shift to nuclear will be the major factor determining whether the world can slow catastrophic climate change during the next half century. In that sense, the Indian shift to thorium power would be a world-saving development. Based on a possible nuclear war with Pakistan, though, India's nuclear industry is even more bound by military priorities than that of the United States. India has never signed the Nuclear Nonproliferation Treaty (NPT), which took effect in 1970. Until 2006, India was barred from international trade in nuclear technology. A landmark agreement signed in 2006 with the United States under George W. Bush opened up India to nuclear markets; now suitors from nuclear suppliers in Japan, France, Korea, Russia, and the United States are vying to build India's next generation of reactors. Moving to thorium could make India one of the few truly energy-self-reliant countries in the world and would give a huge boost to the prestige and influence of the roiling, at times ungovernable, Indian subcontinent. India, like China, wants to be the nuclear superpower of the twenty-first century. That's not an impossible goal, but it's a complicated one. And it would be an astounding achievement for a country where more than one-quarter of adults are illiterate, more than one in five lives in poverty, and fewer than 10 percent graduate from high school.

TH90 • TH90 • TH90

THE BHABHA ATOMIC RESEARCH CENTRE was founded in 1954 as the Atomic Energy Establishment and renamed in 1966 for Homi Bhabha, the father of India's nuclear program. Born under the raj to a wealthy family in Mumbai, Bhabha was one of the prodigies who occasionally sprang from British India and landed in the pantheon of British science. (Another prominent example of such a prodigy was the mathematician Srinavasa Ramanujan, who was 12 years Bhabha's senior and who navigated a similar path despite coming from far less fortunate circumstances.) Bhabha attended India's finest schools: the Cathedral Grammar School in Bombay, Elphinstone College, and the Royal Institute of Science. At age 18 he won acceptance at Caius College in Cambridge, where he became one of the top students in math and physics. Eager to see his son acquire a practical degree that would enable him to become one of those building a modernized India, Bhabha's father obliged Homi to pass the "Tripos" exam in mechanical sciences, in which he took a first-class degree in 1930. Then Bhabha was able to pursue his more esoteric interests, namely theoretical physics.

Anyone studying physics at Cambridge between the wars had the profound good luck to work under three of the great physicists of the twentieth century: Paul Dirac, James Chadwick, and, of course, Ernest Rutherford. All three were or would become Nobel laureates (Rutherford won in 1908 for his exploration of radioactivity, Dirac in 1933 for his contributions to quantum mechanics, and Chadwick in 1935 for his discovery

of the neutron). While working at the Cavendish laboratory and traveling abroad to study with Wolfgang Pauli, Enrico Fermi, and Niels Bohr,
Bhabha came up with the first calculations to accurately describe the cross
section of electron-positron scattering, a phenomenon that would come to
be known as Bhabha scattering.

When World War II came, Bhabha found himself in his homeland for
what he thought would be a brief holiday. He never lived abroad full time
again. In relatively short order, aided by the high level of education in the
physical sciences in India (a legacy of the raj), Bhabha forged a nuclear power
research establishment on the Indian subcontinent. He founded the Cosmic
Ray Research Unit at the Indian Institute of Science in Bangalore, the Tata
Institute of Fundamental Research (which, when India won independence
in 1947, moved into the buildings of the former Bombay Yacht Club), and
ultimately the Atomic Energy Establishment at Trombay. Bhabha died when
an Air India flight, on its way from Bombay to New York, crashed into Mont
Blanc in January 1966. The dark state of U.S.-India relations in the Cold War
gave rise to conspiracy theories around the wreck, including speculation that
the CIA wanted Bhabha dead in order to halt the advance of India's nuclear
program.

If so, it didn't work. The world's youngest and most unruly democracy
wanted to join the nuclear club. It saw a deeply hostile military state, Pakistan, on its western border and, to the northeast, Maoist China, which tested
its first atomic weapon at Lop Nur in 1964. Beyond was a developing world
being carved up by the two superpowers, communist and capitalist. India's
Atomic Energy Commission (headed by Homi Bhabha), created by Jawa-
harlal Nehru in 1948, had classified all nuclear R&D as secret and decreed
government ownership of all nuclear materials, including uranium, of which
India had little, and thorium, of which it was known even then to have a great
abundance. Indira Gandhi, the daughter of Nehru, was determined to follow
her father's policies of a quasi-socialist economy, a foreign policy free of superpower domination, and a de facto one-party democracy. From the outset
the Indian nuclear program benefited from eager assistance from outsiders,
including France, Canada, the Soviet Union, and the United States, even as
Indian researchers systematically diverted the technology to weapons building. In late 1955 nearly two million Indians attended an exhibit in Delhi by

the U.S. Agency for International Development that featured explanations of atomic science, including working models of nuclear reactors.

Thanks to Cold War strategic imperatives, as well as the charismatic Bhabha's relentless lobbying, the United States became India's leading supplier of nuclear technology and material. India's leaders played the two superpowers off each other expertly. From four American presidents New Delhi wangled more than $93 million in loans and grants for nuclear power. Most of the money went to the construction and operation of India's first power reactor, at Tarapur. But plenty of the spillover went into bomb research.[1]

Less than 20 years later India—which, along with Pakistan, North Korea, and Israel, had refused to sign the NPT since its 1968 approval—exploded its first nuclear device, a bomb code-named Smiling Buddha, in the far western Thar Desert. Despite the extensive nuclear aid provided by the United States for two decades, and despite a sizable collection of CIA reporting that left little doubt that India was building a bomb, U.S. policy makers managed to be surprised at this development. The Smiling Buddha detonation ensured India's status as a nuclear pariah for 34 years, but it did not prevent the government from tirelessly pursuing nuclear power as the country's eventual primary source of energy. By some estimates, since the early 1970s India has spent 20 to 25 percent of all its scientific R&D funds on nuclear power.

It has not been a profitable investment. Proceeding largely without imports or international technical collaboration, India's nuclear power industry is among the poorest performing in the world. The industry is chronically short of uranium—though thorium is plentiful, U-238 deposits in India total less than 86,000 tons, and its 19 conventional light-water reactors run at about half their capacity. Imports, mostly from Russia, are expensive. The grid is shoddy; a 2007 report by the consulting firm KPMG found that Indian utilities lose 30 to 40 percent of generated power in transmission and distribution, costing about $6 billion a year.[2] Rajasthan 1 and 2, advanced heavy-water reactors built with Canadian assistance under an agreement signed before Smiling Buddha was detonated, have been beset with technical difficulties; Rajasthan 1 has been shut down since 2004 as the government ponders its future. In the 1960s Homi Bhabha predicted that by 1987 India would produce 18,000 to 20,000 megawatts of nuclear power; the real figure turned out to be 512 megawatts, less than 3 percent of Bhabha's forecast. The country pursuing thorium power most vigorously has an operational record in nuclear energy that is hardly encouraging.

Undeterred, the scientists at BARC are pursuing the "three-stage program" first laid out in the 1950s by Homi Bhabha himself. It's hard to have a conversation about nuclear power with an Indian without hearing about the three-stage program. In industry jargon it's a PHWR-FBR-AHWR (pressurized heavy-water reactor–fast breeder reactor–advanced heavy-water reactor) progression of facilities. First, plutonium is produced in pressurized heavy-water reactors (which use mainly natural U-238 but require some enriched uranium as well). In stage 2, already underway, the plutonium is used in fast breeder reactors to breed U-233 from thorium as well as additional plutonium from uranium. Finally, big advanced heavy-water reactors will use thorium and U-233 to supply the bulk of India's power needs in the future. These three stages will turn India into the world's first thorium nuclear country. Recognizing that India's uranium resources were scarce, Bhabha wished to build a thorium-based nuclear power industry that would make India self-sufficient in energy. That the three-stage program would produce plutonium, suitable for weapons, was a welcome side benefit. Bhabha's program is an elegant, expensive, and needlessly complicated scheme. To question it is to throw dust on his legacy.

Already, construction is under way at Kalpakkam on the southeast coast near India's second major nuclear facility, the Indira Gandhi Atomic Research Centre, on a 500 megawatt fast breeder reactor of the sort that will burn plutonium and thorium in stage 2. Designs for stage 3, advanced heavy-water reactors, are thought to be far along, although India remains secretive about its nuclear technology (the country's Department of Atomic Energy turned down multiple requests to visit BARC in my research for this book). That's not to say that Indian leaders haven't been vocal about their ambitions.

"The world must wake up" and support India's thorium power program, declared Srikumar Banerjee, chair of India's Atomic Energy Commission, at an international conference on nuclear power in Paris in March 2010. International estimates of future demand for nuclear energy sharply underestimate the most likely scenarios, Banerjee said; they're based on First World growth rates of 1 to 2 percent a year in energy demand, while power requirements in the Third World are increasing by more than 10 percent a year. Supplying

India with its burgeoning power requirements would mean burning more than five billion tons of coal a year, four times the current coal consumption in the United States, producing 8.5 gigatons of CO_2 emissions—a pollution load that "would be shared by all of you," he added, in an explicit warning to the developed nations of the West.[3]

At the levels of consumption projected by Indian officials, the world risks running out of uranium sometime after 2050 (a claim, it should be noted, at which Western nuclear executives scoff). Only thorium can provide a sustainable source of energy beyond the twenty-first century. "Unless thorium is used, it will not be possible to have sustainable energy on this planet," Banerjee warned. "Unless we wake up, humans won't be able to exist beyond this century."[4]

TH90 · TH90 · TH90

TWO TYPES OF OBSTACLES STAND IN THE WAY of India's three-stage thorium power program: political and technical.

The political barriers were underlined in June 2011 when the 46-nation Nuclear Suppliers Group (NSG) moved to enact more stringent technology-transfer rules for countries like India that have never signed the NPT. The move was as unexpected as it was overdue. The new guidelines—which, to be sure, are nonbinding—would bar non-NPT nations from access to the core technologies for enrichment and reprocessing, which India, despite its aspirations to self-sufficiency in nuclear power, badly needs to create a uranium-plutonium-thorium fuel cycle.

India, though, occupies a special place among nuclear nations, thanks to the landmark nuclear deal signed with the Bush administration in 2006.

That deal gave India a "clean waiver" that exempted it from restrictions imposed on other non-NPT countries. In other words, the deal made India the only nation in the world allowed to freely pursue civilian nuclear commerce without signing the NPT. Since the NSG announced its new technology-transfer rules, several nuclear powers, including the United States, France, and Russia, have thrown their support behind India, effectively rendering the NSG rules moot.

Nevertheless Indian officials professed outrage. Anil Kakodkar, the former director of BARC, called the NSG rules a betrayal. The NSG move "essentially targets India as we are the only country outside the NPT eligible for nuclear transfers."[5]

This is not a totally specious claim. The suppliers group has long pursued policies that appear to be strongly opposed to the proliferation of nuclear weapons but in fact accommodate the existing nuclear trade of the suppliers themselves. Indian officials argue that a closed fuel cycle—one that leaves behind little to no toxic waste—based on thorium offers safeguards against proliferation that no international regulations, of the sort promulgated by the NSG, can equal. Whether solid fuel thorium reactors, like the ones that India plans to build, would contribute to proliferation is a matter of some controversy. But standing in the way of India's thorium power plans will not necessarily reduce the chance that weaponizable material will go astray. The advantages of thorium, which lessen the risk of proliferation, should lead to more collaboration on the three-stage program, not less.

That's international politics; the domestic political landscape is equally sown with land mines. The Department of Atomic Energy has signed a series of agreements, largely with French and Russian suppliers, to build conventional reactors, including two 1,000-megawatt Russian machines at Koodankulam, in the southern state of Tamil Nadu, that have sparked fierce protests. Opponents include not only those who oppose nuclear power on principle but also those who charge that greedy politicians are abandoning Homi Bhabha's sacred three-stage program to appease foreign powers and line their own pockets. The government "has quietly sought to abandon the Three Stage Programme in favour of a massive programme of purchasing foreign reactors that give zero benefit to local technology and very little to local industry," claimed a report in the opposition *Organiser* newspaper. "The US, China and the EU are using every means of pressure at their disposal to prevent India from mastering the Fast Breeder Reactor technology, because they know that once such a Rubicon gets passed, India would become one of the key countries in international nuclear commerce."[6]

Such charges are common currency in India's fractious political market. They were given added weight by the revelation, in documents released by WikiLeaks in early 2011, that the Singh government had purchased votes in parliament to win approval of the 2006 nuclear deal with the United States. The ensuing uproar—coming on the heels of a similar scandal involving the granting of telecom licenses for India's exploding mobile phone market—promised to at least delay the lucrative deals with foreign nuclear power contractors.

Then there are those who simply believe adding nuclear power after Fukushima is morally wrong. With its strong heritage of Gandhian nonviolent protest, India has a nascent but vocal antinuclear community of activists, represented by groups like South Asians Against Nukes and the Indian Social Action Forum. Much of this opposition has centered on the Koodankulam plant as well as the plan for three massive Areva-built "evolutionary pressurized reactors" near Jaitapur, a port city on the Arabian Sea. Opposition in Jaitapur, a region of rich cashew fields, mango orchards, and fishing grounds, has expanded rapidly and violently. In April 2011 one person died when police fired on a crowd of protesters who had gathered in opposition to the project. A week later—on the twenty-fifth anniversary of the Chernobyl disaster, as it happened—Prime Minister Manmohan Singh held a high-level meeting with the chief minister of Maharashtra state and several cabinet members, including the minister of environment and forests and the secretary of the Department of Atomic Energy, to review the status of the proposed Jaitapur Nuclear Power Plant. This was at a time when the full dimensions of the Fukushima nuclear accident were still being revealed. "However, instead of conducting a meaningful review, the Government simply held a press conference and reiterated its determination to go ahead with the project," reported Suvrat Raju and M. V. Ramana, physicists with the Coalition for Nuclear Disarmament and Peace.[7]

Conspiracy theories, mass protests, and vote-buying scandals aside, though, it's likely that India—with a potential nuclear market of tens of billions of dollars—will find a way to obtain the technology it needs to push ahead with the three-stage program. The larger obstacles lie in the technology itself.

TH90 • TH90 • TH90

THE FIRST TECHNICAL HURDLE IS INDIA'S SPOTTY RECORD of operating nuclear power plants. India actually has a respectable nuclear safety record. But this, after all, is the country of the 1984 chemical disaster at the Union Carbide facility at Bhopal, which killed as many as 11,000 people and injured hundreds of thousands more. And the behavior of the Indian government since Fukushima has not inspired confidence. Unlike many nuclear powers, which undertook comprehensive reviews of nuclear safeguards in the wake of the Japanese disaster, New Delhi has responded largely with a public relations campaign to convince the public that Indian nuclear plants are safe and secure, a campaign based mostly on repeating assertions that they are so.

Like many countries, India has aging nuclear plants that have been patched and upgraded in order to extend the reactors' working life. The reactors at Tarapur, India's largest nuclear plant, include two boiling water reactors (the same design as the Fukushima machines) that have been licensed to continue running for decades beyond their projected lifespan. Spent fuel rods from the plant, accumulated over four decades, are kept in pools without any containment structure—an arrangement that represents "safety and environmental hazards that are greater than at any other plant in the world."[8]

In 2010 a worker at a scrap metal yard died from radiation exposure after coming into contact with a piece of machinery, apparently discarded by the chemistry department at Delhi University, that contained deadly cobalt-60. Several containers of Indian steel have been turned away from European ports in recent years after testing for high levels of radiation. The incidents have raised concerns about the handling of nuclear materials in the country.

Anyone who has ever traveled in India will be familiar with the extreme polarization of its society and economy: legless indigents begging outside modern skyscrapers, a growing educated middle class alongside surging populations of slum dwellers, business people with the latest mobile gadgets stuck in traffic behind oxcarts from the eighteenth century, and so on. Nearly 36 percent of India's population has no access to electricity, and explosive economic growth and continuing population increases ensure that the energy sector will forever be playing catch-up. For the ruling party this brings enormous incentives to cut corners and go easy on safety enforcement at a time when it is signing multibillion-dollar deals with foreign corporations to dramatically expand the nuclear power industry.

The real question that shadows India's three-stage program is whether the political system has become too corrupt and corrosive to effectively promote nuclear safety in the long run. Building dozens of new reactors is a daunting challenge in any country; in India it's made more so by a political culture that remains rooted in graft, cronyism, and the distribution of favors, including lucrative infrastructure contracts, to loyal supporters.

What's more, the three-stage program itself involves a tricky path to safe, clean, thorium-powered nuclear plants. The second stage, now under construction, involves fast breeder reactors. Fast breeders have proven perilous even for the most technologically sophisticated nuclear nations in the world; no country has ever succeeded at building fast reactors at anything like the scale Indian leaders envision. Suffice it to say that India's unique heritage and special circumstances make it a less-than-outstanding candidate to be the first.

The history of India's first fast breeder test reactor, on which construction began at the Indira Gandhi Atomic Research Centre in the 1970s, has been a parade of accidents, shutdowns, and do-overs; the reactor went critical only in 1985 (9 years after the planned criticality date) and began operating a steam generator in 1993, 22 years after the project began. It took 15 years before the reactor achieved 50 days of continuous operation at full power, and in its first two decades the reactor ran for a total of about 36,000 hours—which means it actually produced power about 20 percent of the time.[9]

Efficiency is one thing; safety is another. There is evidence that the Department of Atomic Energy and the scientists working on the fast breeder program have downplayed the danger of fast breeders while scrimping on the rupees required to make India's reactors safer. To understand these dangers, it's worth briefly reviewing some factors that make fast breeders such twitchy beasts.

Most fast breeder reactors have a "positive coolant void coefficient." That is a nuclear engineer's way of saying that when voids (usually steam bubbles) form in the core, the rate of reactions increases and the energy level rises. When accidents occur in these reactors, such voids tend to form, making them harder to control and more prone to melt down. If enough bubbles form, and the coefficient is large enough, all the coolant in the reactors can quickly boil, leading to disaster. Reactors with positive void coefficients are illegal in the United States.

India is also planning fast breeder plants with relatively weak containment plants. "Both of these design choices—a weak containment building and a reactor core with a large and positive void coefficient—are readily explainable," Raju and Ramana write. "They lowered costs."[10]

The other reason to mistrust the three-stage program has to do with nuclear proliferation—a paradoxical fact since one of thorium's primary advantages is that it reduces the risk of weaponizable material reaching the wrong hands. One of the program's key components is the production of large amounts of plutonium. The prototype fast breeder alone requires almost two tons of plutonium in its core at start-up. Success for the three-stage program would significantly multiply the amount of plutonium in the world. India has refused to publicly pledge that its reprocessing plants and its stockpile of plutonium will be used for civilian purposes only; plutonium from reprocessed uranium and future breeder reactors could be enriched for atomic weapons. India has every intention of continuing to produce plutonium for its own nuclear arms. Given the state of security on the Indian subcontinent, it's hardly a leap to conclude that large amounts of plutonium being produced at large nuclear plants around the country represent tempting targets for those wishing to build illicit weapons of mass destruction. While nuclear vendors from the West rush to sign contracts for the three-stage program, the risks of proliferation rise.

TH90 • TH90 • TH90

GIVEN ALL THIS, I HAD TO ASK, why bother? Blessed with large thorium reserves and an existing nuclear R&D capacity that, operational snafus notwithstanding, is world class, India, rather than taking a laborious three-stage route to thorium-based nuclear power, could start building thorium reactors—most simply and inexpensively, liquid fluoride thorium reactors—tomorrow. The reasons it's not doing so have to do with institutional inertia, national pride, and supposed national security concerns—such as, for instance, building its nuclear arms stockpile. China, meanwhile, is taking a more catholic approach to its nuclear power program, including investigating LFTRs.

In a development heralded by thorium advocates around the world, China officially announced in February 2011 at a Shanghai scientific conference that it will begin a program to develop a thorium-fueled molten salt reactor (MSR), aka an LFTR. The project was first reported on the mainland in the *Wen Hui Bao* newspaper. I broke the news in the West in a story for Wired.com. I first heard about it at a conference in Oak Ridge with Sorensen and other thorium activists. The phrase "*Sputnik* moment" was used freely.

The world's most dynamic economy had thrown down the thorium gauntlet. While India chose to slog up the long hill of its three-stage program, China was going straight for the prize.

India's three-stage program calls for gradually phasing in thorium fuel rods in advanced heavy-water reactors. The Chinese program, in contrast, marks the largest national initiative to pursue thorium MSRs to date. One of the world's largest consumers of coal for electricity, the People's Republic has embarked on a public campaign to shift toward less noxious energy sources, including nuclear power. The massive Three Gorges dam project, one of the largest public works projects in history, was designed to produce 18.2 gigawatts of electricity and has also engendered fierce criticism and internal protest. Electricity demand is growing at nearly 10 percent a year, and Chinese officials, often willing to ignore international objections to its domestic policies, are committed to using nuclear power as a source of clean, inexpensive energy.

The nuclear ambitions of India and China are similarly outsized, but the cultures and capabilities of the two countries are quite different. I used to live in Hong Kong, and I've traveled extensively in both northern India and southeastern China. The differences in the countries, for me, can be summed up with a glance at their railways: The Indian rail system, a source of national pride since the days of the raj, is known neither for its modernity nor its efficiency. In September 2011 the passengers on a cross-country journey were surprised to learn that their train had somehow traveled more than 600 miles in the wrong direction. This was treated as a newsworthy but not completely unheard-of experience. The passengers, suitably outraged, stormed the depot.

In China the government completed the Beijing-to-Tibet railway in 2006, a dream since the days of Sun Yat-sen. Totaling 2,526 miles, it includes tracks, from Golmud to Lhasa, at the highest altitude of any railway in the world. The two-day journey, which passes through the world's highest-altitude railway tunnel and uses many sections of elevated track passing over permafrost, costs about $160, or about what it costs to go from Boston to Washington, D.C., on the relatively low-tech Acela train. The new Chinese line has engendered plenty of criticism regarding fears of cultural hegemony and the loss of Tibetan autonomy, but no reports of wrong-way trains have surfaced. In the realm of public infrastructure, India is a great producer of think-tank studies, government reports, and beard-stroking orations. China, unimpeded by the hurly-burly of parliamentary democracy, is a better place for actually accomplishing things. If you are betting on which country will build a thorium power reactor first, the choice is not tough. (A July 2011 crash on a high-speed rail line near Wenzhou, on the southern coast, killed 39 people and sparked a level of public outcry seldom seen under communist rule on the mainland. In public statements after the accident, Chinese premier Wen Jibao vowed to toughen safety standards in China's rapid industrialization—but the crash did little to slow China's drive to modernize its energy and transportation infrastructure.)

China has 14 nuclear power reactors in operation on the mainland today, with more than 25 under construction and more soon to get under way. For many years a consumer of reactor technology and components from the West, and from Russia, China will soon be building fully homegrown reactors. The development of liquid fluoride thorium reactors would make China the most advanced nuclear power nation on Earth—and could well give it yet another source of high-tech products to pad its export surplus.

Comparing nuclear reactors to humble kitchen appliances, Xu Hongjie, a research scientist at the Shanghai Institute of Applied Physics, said, "We need a better stove that can burn more fuel."[11] It was a line reminiscent of Chairman Mao's finest exhortations.

Like many nuclear nations, China declared a pause to review and reassess its nuclear development plans after Fukushima. This was only a breather; Chinese officials made it clear that the Japanese accident would not affect their long-range plans. And they scoffed at the German decision to get out of nuclear power altogether. The comments of Chinese officials did not inspire confidence. Dr. Liu Changxin, vice general secretary of the China Nuclear Society, remarked that such natural disasters "don't happen in China"—a startling claim given the devastation wrought by the 2008 earthquake in Sichuan Province, which killed 69,000 people and left nearly five million homeless.

The Chinese thorium program is headed by Jiang Mianheng, an electrical engineer and the son of the former Chinese president Jiang Zemin (see chapter 1). Jiang Mianheng, who is also a vice president of the Chinese Academy of Sciences, headed a Chinese delegation that visited Oak Ridge in the fall of 2010. The Chinese politely listened to the research presentations, and patiently endured the facilities tour, before revealing that what they were really there for was to soak up as much information on thorium MSRs as they could. "They were quite open about it," a person present at those discussions told me. In early 2012 Western observers of the Chinese nuclear effort stated that the Shanghai Institute of Applied Physics, with around 400 people and a budget of $400 million, planned to build two prototype molten salt reactors by 2015.

Like India, China needs to shift to nuclear from coal to avoid adding catastrophic levels of carbon to the atmosphere. At the same time many in the U.S. thorium movement regard the development of Chinese LFTRs as a direct threat to U.S. economic competitiveness. The specter of Chinese competitiveness with the United States is often overblown; in general, China's prosperity and the well-being of its people, are good things for the world, particularly for Americans. That won't make it feel any better when we are buying LFTRs with "Made in Shanghai" stamped on the side.

The alarmist version of China's next-generation nuclear strategy comes down to this: if you like foreign oil dependency, you're going to love foreign nuclear dependency.

While various international efforts, including the Gen IV nuclear R&D initiative, include a thorium MSR component, China has made clear its intention to go it alone. The announcement from the Chinese Academy of Sciences states explicitly that the People's Republic plans to develop and control intellectual property with regard to thorium for its own benefit. "This will enable China to firmly grasp the lifeline of energy in its own hands," *Wen Hui Bao* reported.[12]

The plans for China's lifeline include not only thorium but also critical materials that have increased in value at a startling rate since 2010 and of which China now has a monopoly: rare earth elements.

TH90 • TH90 • TH90

NOT LONG AGO, FEW AMERICAN POLITICIANS or journalists could have told you what "rare earth" is. Now they are at the center of a controversy that includes issues of national security, trade policy, green energy, and high-tech supremacy. The "rare earths crisis," as it's now termed, has been covered at length in the *Wall Street Journal, New York Times, Financial Times,* and *National Geographic.*[13] To thorium advocates, though, the critical importance of these unique elements is not news.

The term "rare earth" is a misnomer: rare earth elements are actually quite plentiful. Comprising the lanthanoid elements—lanthanum (from the Greek *lanthanein,* to be hidden) through lutetium, numbers 57 to 71 on the periodic table—plus scandium and yttrium—they are most often found in the ground in association with monazite, the reddish-brown ore that also happens to be the principal source of thorium. When you find thorium, you usually find rare earth elements. Where you find rare earths, you almost always find thorium.

They are rare in the sense that, until relatively recently, they were exceedingly difficult to isolate and thus of little or no industrial value. During the Manhattan Project chemists developed the process of ion exchange for separating and purifying rare earths. Of course, they were actually working on purifying plutonium-239 for bombs; the separation of rare earths was a happy by-product of that research. Rare earth elements became valuable in the 1950s, with the advent of color TVs: they were (and still are) used as phosphors to brighten the colors of television tubes.

Rare earths are still complicated and costly to extract today. Mining gold, for example, is much simpler than mining rare earth elements. Significant deposits of rare earths are found in Australia, Brazil, India, Malaysia, South Africa, Sri Lanka, Thailand, and the United States, and for many years the world's leading producer of rare earths was, as you might imagine, the United States. Through the 1990s the Mountain Pass mine in California was the world's largest source of rare earths. Then, as low-cost Chinese-produced rare earth elements became available, Americans lost interest in actually finding and producing the stuff themselves. China now controls about 97 percent of the world market for rare earths.[14]

That fact is largely the doing of one man, Xu Guangxian, who is the inventor of separation techniques for both uranium isotopes and rare earths in China. Like many Chinese scientists, Xu was trained in the United States, at Columbia University, where he received his doctoral degree in chemistry in 1951. He pioneered the extraction of fissile fuels before being caught up in the Cultural Revolution, when he and his wife were sent to a labor camp. Released in 1972, he returned to Beijing University and began to apply his previous work on uranium to rare earths. He was successful. Chinese leaders were not slow to realize the strategic importance of a domestic rare earths industry. Chinese premier Deng Xiaoping reportedly declared, "There is oil in the Middle East; there is rare earth in China." In 2009, at the age of 89, Xu was awarded the State Supreme Science and Technology prize, the Chinese equivalent of the Nobel.

As Chinese companies began flooding the market with inexpensively produced rare earths in the 1980s, the price began to fall. Between 1992 and 1996 the price of a ton of rare earths fell from $11,700 to $7,430, and undersold U.S. producers started getting out. Rare earths production in the United States effectively ceased in the early 2000s.

Today these elements form essential components in much of the advanced technology used in communications, weapons systems, and even oil refining. They have been designated strategic materials by both China and the United States. They're found in missile guidance and radar systems, smartphones, laptop and tablet computers, as well as such green technology as next-generation wind turbines and hybrid vehicles. And because rare earths occur in such close association with thorium, China—which for many years simply set aside thorium-bearing ore in tailings after removing the rare earths—now has an abundant supply of thorium.

The Chinese takeover of the rare earths industry is epitomized by what happened in Anderson, Indiana in the mid-1990s. A company called Magnequench, a division of GM that made sophisticated magnets from iron, boron, and the rare earth neodymium, was sold to a Chinese consortium allied with a U.S. investment firm, the Sextant Group, founded by Archibald Cox Jr. (who had gained fame as the special prosecutor in the Watergate affair). The Chinese group won approval for the buyout by agreeing to keep the company in the United States for five years. The day after that deal expired in 2002, "the entire operation, along with all the equipment, disappeared. . . . At the time, it seemed that no one really cared."[15]

Today those magnets are used in critical military systems, including lasers, range finders, traveling wave tubes, and Klystrons, which are used in satellite communications. The United States has no domestic capacity to produce the magnets.

As I learned about the death of the rare earths industry in the United States, I couldn't help thinking about the abandonment of thorium and molten salt reactors. Both technologies were pioneered by Americans (or scientists working in the United States); both were based on elements that are abundantly available in U.S. soil; and both have become strategically important in the twenty-first century for energy security and national security. The difference is that thorium power was consciously set aside in favor of an existing and established technology. The rare earths industry was more a case of negligence: with nobody really paying attention, U.S. producers simply abandoned it when it became less profitable. It was as if the United States could no longer be bothered to find and develop the materials on which its high-tech economy and national defense depend.

TH90 • TH90 • TH90

STARTING IN 2007, CHINA DRAMATICALLY CUT export quotas for rare earth elements, capping foreign shipments at 7,976 metric tons in the last six months of 2010, down from 28,417 tons for the same period in 2009. In response, the World Trade Organization ruled that the export curb violated international trade agreements and demanded that China return to full production. Chinese leaders were defiant. Interviewed by the official newspaper, *China Daily*, Huang Dongli, a researcher at the Institute of International Law, which is affiliated with the Chinese Academy of Social Sciences, said the cuts were justified based on "conservation of exhaustible natural resources."[16]

"Rare earths are nonrenewable resources of strategic importance," declared deputy commerce minister Zhong Shan, speaking at a conference on rare earths in Baotou, in Inner Mongolia, in July 2011. Hosting a conference in Inner Mongolia pretty much guarantees that you'll have not only the market but the dais to yourself. China's export restrictions, Zhong said, "will help the country protect the environment and accelerate the industry's restructuring."[17]

In fact, Chinese officials have said that by 2015, they may export no rare earths at all, because domestic demand, growing quickly as China becomes the world's major source of many high-tech components, will consume all the production from its mines. By the fall of 2011, Chinese mining companies were reportedly trying to buy up rare earth sources outside the mainland.[18]

Given this situation, it was inevitable that U.S. companies would try to reenter the market. And since the association of rare earth elements and thorium is so strong, it was inevitable that the thorium movement would align with those who believe that domestic rare earth production is critical to the future economic competitiveness of the United States. That started happening in early 2011.

Jim Kennedy, a St. Louis–based developer who has become the most vocal figure demanding the return of a domestic rare earths industry in the United States, began an effort, along with the Thorium Energy Alliance, to reopen the Pea Ridge Mine in Missouri. Kennedy actually has a permit for the Pea Ridge Mine, but there's a problem: because the rare earths are interlaced with thorium—which is, after all, a radioactive element—the regulations for handling the ore are highly restrictive, and the processing is costly. Wings Enterprises, Kennedy's company, has reached agreements with private funders to reopen the Pea Ridge Mine—but there are no facilities to process the ore. Like many self-made men, Kennedy was not to be deterred. Unlike many such men, he came up with an audacious solution: have the government step in.

Under legislation formulated by Kennedy and John Kutsch, head of the Thorium Energy Alliance, and hand-carried to Capitol Hill, Congress would authorize and provide loan guarantees for a rare earths processing facility (preferably in Missouri near the Pea Ridge Mine) that will also separate thorium, to be stored until a thorium-based power industry emerges to use it to generate electricity. This was free-market economics that a central planner could get behind. Congress had effectively outlawed the processing of rare earth elements because of the thorium problem; Kennedy and Kutsch wanted Congress to solve the problem by creating a repository for rare earths and thorium. Kutsch compared it with a grain elevator for rare earth and thorium producers; instead of processing and storing their own grains, farmers pool their funds, get a little help from the feds, and build a centralized facility. So it would be for miners of rare earths and thorium. China would lose its rare earth monopoly, the United States would jump-start a domestic thorium power industry, and no longer would you have to travel to Inner Mongolia to attend a rare earths conference.

Even for a thorium advocate like me, that sounded like a rosy scenario. Depending on Congress to solve the viciously complicated dual challenges of rare earth elements and thorium seems like a fool's errand.

TH90 • TH90 • TH90

THERE WAS ONE OTHER FLAW in Kennedy and Kutsch's plan to solve the rare earths extraction problem and the thorium puzzle in one legislative swoop: an American company already is planning to produce and process its own rare earth elements—without government assistance, thanks very much. The Denver-based Molycorp owns the Mountain Pass mine, which for many years was the world's premier rare earth mine.

Discovered in 1949 by, appropriately enough, a uranium prospector who noticed the high radioactivity of the surrounding rocks, Mountain Pass is set in a windswept gorge on the landward side of the Clark Mountains in southeastern California. First developed by the Molybdenum Corporation of America (predecessor of today's Molycorp), it has been producing rare earths since the early 1950s.

Beginning in the 1990s, a combination of overseas competition and local environmental opposition doomed the mine. After producing rare earth elements for a half century, Mountain Pass was finally shut down in 2002, after its then-owner, Unocal, paid more than $1.4 million in fines and settlements because of environmental damage from radioactive tailings. A federal investigation concluded that a pipeline carrying radioactive water—contaminated with thorium, which was considered useless—to evaporation ponds at Ivanpah Dry Lake near the Nevada state line had leaked into the watershed. Meanwhile low-cost Chinese production had undercut the price of rare earth from Mountain Pass, making the fight with state authorities no longer worth the candle.

A far-sighted group of investors purchased the mine in 2008 and hired long-time Unocal executive Mark Smith—who has spent most of his career working on Mountain Pass, for one company or another—to run it. Molycorp went public in July 2010, placing itself at the center of the rare earths frenzy, and became the most successful IPO of the year. Its share price nearly quadrupled between February and July 2011. Unlike Kennedy, Smith has no plans to capture and store the thorium by-product.

The ore at Mountain Pass contains lighter rare earth elements than the Pea Ridge material, which is richer in both thorium and the heavy rare earth elements, which are expected to be in short supply by 2017 (many of the lighter rare earths are actually in surplus, or soon to be). Because the Mountain Pass rare earths contain less radioactive material, it has been easier for Molycorp to win approval from government agencies. In December 2010 the company said that it had secured all the environmental permits needed to begin construction of a new ore-processing plant at the mine; environmental monitoring and compliance is expected to cost about $2.4 million a year. In

October 2011, the company announced that it had started new production at the mine.

"We want to be environmentally superior, not just compliant," Smith boasted when I talked to him. He also said Molycorp has no plan to use the thorium from Mountain Pass. The thorium from Mountain Pass will be discarded, disposed of in a paste form that will be set in concrete and layered into the landscape at a cost of millions of dollars. "We'd rather put it right back where it was," he said. Once again the energy future is being tossed away as radioactive waste.

TH90 • TH90 • TH90

MEANWHILE, BACK IN CHINA, not everyone has clambered aboard the nuclear bandwagon. For obvious reasons, China lacks a tradition of outspoken environmental activism, but a few voices have begun to question the massive nuclear construction program. In a rare display of public dissent, Professor He Zuoxiu, one of the mainland's leading theoretical physicists, compared the nuclear power plan to Mao's Great Leap Forward, when ill-conceived plans for rapid industrialization and backyard steel smelters resulted in a disastrous famine. China is "seriously unprepared, especially on the safety front," for the rapid nuclear expansion, He wrote. "We have not solved the problems of technology, cost or safety but rashly rushed out an over-ambitious plan. I think it is a mission impossible." Part of the problem is importing sufficient supplies of uranium: "How much can we import from abroad? And would it not be more difficult than importing oil and gas?"[19]

Thorium-based reactors would solve the uranium supply problem. In January 2011 news sources on the mainland reported that China had finished a pilot fuel-reprocessing plant—a facility that can and most likely will be used to reprocess spent uranium into fissile material that can be used to ignite thorium reactors. According to Chinese state television, China National Nuclear Corporation, the state-run nuclear power monopoly, completed the pilot plant at its No. 404 facility in the Gobi Desert in remote Gansu province. If true, the achievement places China in a select group of nuclear powers with the technology and the know-how to recycle irradiated fuel.

Also, like India, the People's Republic has been able to make use of generous partnership agreements with Western nuclear states to further its own technology program. And, as in India, Western vendors have been eager to participate. In 2009 Atomic Energy of Canada, a government agency, signed an agreement with the Nuclear Power Institute of China and two mainland corporations to jointly develop thorium-fueled reactors of the Canadian CANDU design. (CANDU is short for Canada Deuterium Uranium reactor. First developed in the 1960s, CANDU machines are heavy-water–moderated reactors that can run on unenriched uranium—or thorium.)

By 2011, though, the Ottawa government had sold off Atomic Energy of Canada's commercial reactor division to a private engineering group, and all thorium-power R&D in China had been brought in-country, according to thorium activists in the West. In the summer of 2011 the Chinese government declared that it will build a strategic reserve of rare earths—along with, most analysts believe, the thorium that comes with them. China is also stockpiling talent. While precise figures are hard to come by, it's clear that a large percentage of graduate students in nuclear engineering programs in the United States are Chinese nationals. Master's and doctoral degrees in hand, most do not stay in Ann Arbor, Madison, and Cambridge. They go back to the mainland, where there's actually an emerging shortage of trained nuclear scientists and technologists. Recently a Chinese official speaking to a German newspaper jovially invited German nuclear engineers, out of work after the government shuts down that country's reactors, to come work in China. "It is a win-win situation," a Chinese nuclear physicist remarked.[20]

With the assistance of Western vendors and governments, India and China are building nationalist nuclear power industries based on thorium. There's nothing illegal about this. Arguably, hoarding rare earth elements and thorium could be seen as violating international trade agreements. But it's not immoral, or even unethical, especially when there's little market in the West for thorium. In fact, looked at from an internationalist perspective, it will be a very good thing if the world's two fastest-growing large economies wean themselves from coal. Growing, technologically sophisticated nations in South and East Asia that manage to fuel their economies with zero-carbon thorium-based nuclear power will help stabilize both the climate and geopolitics. As citizens of Planet Earth we should all welcome these developments. For Americans, though, from the point of view of energy independence and balance of trade, these developments are ominous. Throughout the second half of the twentieth century, the United States found itself importing oil from Middle Eastern nations that were culturally alien, ruled by despots, and inhabited by millions of people who viewed America as a decadent and evil civilization. In the twenty-first century the United States could find itself importing thorium power technology from India, a chaotic one-party democracy that has never chosen to sign the NPT, and from China, a nominally communist system whose leaders share neither our interests nor our ideals. This is a nationalist point of view on an industry that is thoroughly globalized. The thorium power movement cannot succeed without international cooperation and multinational partnerships. And many Americans and Europeans are working to ensure that the future of energy avoids the clashes of civilizations that have troubled its past.

EIGHT

NUCLEAR'S NEXT GENERATION

Seth Grae was a young New York lawyer, active in the movement opposing the proliferation of nuclear weapons, when he met an aging Israeli scientist preaching a new nuclear gospel. The year was 1991. A young achiever determined to make his mark in the East Coast establishment, Grae held four degrees, including a law degree from American University, a master's in international law, and an MBA from Georgetown. Fresh out of law school, he represented dissident nuclear scientists from the Soviet Union. He helped craft international regulations to control the export of nuclear materials from China and the former Soviet Union, and he consulted with the government of India—which had never signed the Nuclear Nonproliferation Treaty (NPT) and, before the landmark 2006 agreement with the United States, was still a nuclear renegade—on both nuclear arms and nuclear power.

By the early 1990s, though, Grae found himself at loose ends. He "was a bored associate at a Manhattan law firm," according to a *Washington Post* profile, "spending his days on a grab bag of international business clients ranging from video game startups to cement companies."[1] Through his contacts in the arms-control community, he encountered Alvin Radkowsky, an American-born Israeli citizen who'd worked for many years on naval nuclear reactors. A nuclear scientist every bit as accomplished as Alvin Weinberg, Radkowsky had been deeply involved with thorium for more than four decades.

As the chief scientist for the Navy Reactors Branch, Rickover's fiefdom, Radkowsky led the design of the first full-scale commercial nuclear power plant in the United States, at Shippingport, Pennsylvania. He was also a protégé of Edward Teller. The personal history of the father of the H-bomb intersects the thorium story at unexpected angles. Radkowsky studied under Teller at George Washington University in the 1940s. Teller recommended Radkowsky to Rickover, when the latter was building his reactor corps and Radkowsky was a civilian nuclear physicist for the postwar Navy. Radkowsky was among the few members of the *Nautilus* team who did not attend the Oak Ridge school. He didn't need to. He was already one of the foremost U.S. experts on reactor design and the nuclear fuel cycle. At some point, like Wigner and Weinberg, he became fascinated with thorium. When Rickover decided to build a dry-land prototype for the aircraft-carrier reactor project, he picked Radkowsky to head it up. And in Shippingport, Radkowsky saw the opportunity to prove his theories about thorium as an improved nuclear fuel.

The Shippingport Atomic Power Station first went critical in December 1957 and produced energy for the Duquesne Light Company for 25 years. It occupies a unique position in the history of nuclear power. It was considered the first full-scale nuclear power reactor with no military use: all it did was produce energy. For the thorium movement it's like the *Sputnik* space capsule: it proved that something that had never been done could be done. Though it was a light-water reactor, it operated as an experimental breeder reactor, designed to produce more fuel than it consumed. And for several years, beginning in 1977, it ran on a blend of uranium and thorium, transmuting Th-232 to fissile uranium-233. After the reactor was shut down in 1982 and a detailed analysis of the core was run, it was determined that the core contained slightly more fuel than it started with. Shippingport proved that you could use thorium as an inexpensive and safe nuclear fuel in a light-water reactor and that you could breed additional fuel with it. This was not alchemy, but it was close. It's still the only commercial reactor to operate over a long period of time on thorium fuel.

Paradoxically, though, Shippingport's success contributed to the narrowing of the options for the young nuclear power industry. Despite the triumph of the thorium breeder experiment, Shippingport's largest effect was to hasten the triumph of uranium-based LWRs: "The momentum it gave to light-water technology was enough to ward off incursions by competing technologies," Weinberg wrote; its success channeled R&D into "a single line of reactor development" that would lead to the stagnation of the nuclear power industry.[2]

In 1972 Radkowsky, the principal designer of the Shippingport plant and the driving force behind the thorium breeder experiment carried out there, retired from the Navy and emigrated to Israel, where he taught at Tel Aviv University. But he never stopped thinking about thorium. In 1983 Teller, who never publicly recanted his support for nuclear arms, contacted his former student and urged him to restart his thorium work and create a new fuel design that would be less likely to contribute to the proliferation of nuclear weapons. Radkowsky was 68. It took him eight years, but he eventually patented an updated version of Shippingport: a design for a reactor core that combines a "seed" of low-enriched uranium rods, to ignite the fission reaction, and a "blanket" of thorium rods that would breed fissile U-233. In 1992, patents in hand, he founded a U.S. company called Thorium Power Ltd., based outside Washington, D.C., to build thorium reactors and to fulfill the vision he'd first had back in the 1950s: a world run on inexpensive energy from thorium. Radkowsky hired Grae to be his corporate counsel and operations man.

In Radkowsky's seed-and-blanket design, fuel rods of uranium oxide are surrounded by a blanket of outer rods made of thorium and uranium dioxide. The uranium starts a chain reaction that bombards the thorium with neutrons; some thorium atoms capture those neutrons, converting the thorium to protactinium, which spontaneously, instantaneously decays into U-233—prolonging the reaction and producing heat and, in turn, electricity. In theory it's a reactor that will need refueling only every 30 years or so and will produce only minimal amounts of toxic waste.

Conventional uranium reactors typically produce 50 to 60 megawatts per kilogram of fuel; thorium-based plants can achieve a rate of 100 megawatts per kilogram, using up a much larger fraction of the energy available in a given unit of volume. The melting point of thorium is about 500 degrees Celsius higher than that of uranium, and thorium reactors—even solid-core ones—are inherently safer. Finally, the core has a longer life: a thorium blanket can sustain a reaction for many years, while uranium rods must be changed out every three years or so.

"Alvin's vision was to advance thorium-based fuels, based on earlier concepts used at Shippingport, with new designs that could be used in the current generation of nuclear power plants," Grae told me. He is now the CEO of Lightbridge, the new name adopted by Thorium Power in 2009. This unlikely duo—an aged American Israeli scientist and a young lawyer—set out to change the power industry based on solid fuel thorium technology.

Radkowsky died in 2005, a year before Weinberg's death. Thorium Power went public in 2002. On its advisory board is Hans Blix, the former director general of the International Atomic Energy Agency (IAEA) and the chief U.N. weapons inspector for Iraq from 2000 to 2003 (he's best known to Americans for contradicting the Bush administration's claims about Saddam's weapons of mass destruction). The chair of the board is Thomas Graham, a long-time nuclear arms diplomat who served as President Bill Clinton's special representative for arms control. Company officials are fond of the automotive analogy: "We're just trying to replace leaded fuel with unleaded," Grae likes to say. "You don't have to replace your car engine or build new gas stations. You just replace the fuel."

TH90 • TH90 • TH90

WHEN I FIRST STARTED REPORTING on the thorium movement—in other words, before I'd heard the full LFTR gospel from Kirk Sorensen—I figured that Lightbridge, née Thorium Power, would be the story. It didn't turn out that way. In fact, according to the true thorium believers, Lightbridge is something of a bastard child, a halfway measure that would neither rid the industry of the evils of uranium nor fulfill the true promise of thorium. As it turns out, Lightbridge itself has moved in other directions.

Certainly things looked promising at the outset. The company's timing was ideal: in the early 1990s, with the collapse of the Soviet Union, a whole class of nuclear scientists and technicians were basically tossed out on the street, with little means of support and plenty of incentive to traffic both their skills and any illicit wares they might have stashed away—a situation that the U.S. government, for obvious reasons, found unacceptable. Thus was born the Initiatives for Proliferation Prevention, which amounted to a make-work program for out-of-work Soviet weapons scientists. For several years in the mid-1990s thorium power benefited from federal grants to conduct research at the Kurchatov Institute—named for Igor Kurchatov, who led Stalin's weapons program, and often referred to as the Los Alamos of Russia. In the spring of 2009 Lightbridge signed a contract with Kras-naya Zvezda (Red Star), a government-owned nuclear design agency, to carry on testing and commercialization of the thorium fuel cycle. Thus, a U.S. company founded by an Israeli citizen is trying to develop the nuclear fuel of the future at a Cold War institute in Putin's Russia. So proceeds the globalization of nuclear power.

In March 2009 Lightbridge researchers for the first time successfully tested one-meter rods in the experimental reactor. Commercial rods are typi-cally 3.5 meters long; Lightbridge would still have to scale up to that length to run its fuel in conventional reactors.

And then there's the waste. And here things get complicated, as Andrey Mushakov, a Russian economist and head of Lightbridge's international op-erations, explained to me at a whiteboard in the company's McLean, Virginia, headquarters outside Washington, D.C. Spent fuel from solid fuel thorium reactors contains both uranium and plutonium, and it's highly radioactive. However, the volume of fuel is reduced by about half, and it has a far shorter half-life: it takes spent thorium fuel about 800 years to decay to an environ-mentally safe level of radioactivity, compared with 10,000 years with used rods from conventional reactors. You could reprocess the spent fuel into fuel for more reactors, although that's not an avenue that Lightbridge is pursuing, for now.

And if the seeds were made of plutonium from old nuclear weapons (or from nuclear waste from civilian reactors) instead of uranium, then thorium could theoretically create electricity while disposing of old nuclear weapons.

You could also, in theory, refine used thorium fuel into weapons-grade material. In fact, spent fuel from a Radkowsky reactor is actually hotter, radioactivity-wise, than conventional uranium fuel. That makes handling it (for would-be bomb makers or for legitimate stewards) more tricky.

But the uranium and plutonium in spent thorium fuel is *denatured,* that is, it's contaminated with useless isotopes, making it much more difficult to wring bomb fuel from it. If you're a bomb maker, you'd be better off just buy-ing natural uranium on the open market and building an enrichment facility.

That's the technology. The business side is less promising: traded over the counter, Thorium Power stock languished below one dollar for many months in 2008–2009. Then its board reduced the number of shares outstanding and changed the company's name to Lightbridge. With those financial moves came a shift in business direction. Since then, Grae has developed a lucrative business in consulting gigs, such as the one with the United Arab Emirates (UAE) that brought in $6.4 million in revenue in 2010.

These days Grae spends most of his time in Abu Dhabi; he would call me from his offices there for our phone conversations. The UAE has become a testbed for the nuclear renaissance, and it might become a model for how nuclear power can be done right. Blessed with about a hundred billion barrels of oil, the UAE government has decided it would rather export the oil to more petroleum-hungry countries and get its own energy from sustainable sources, like solar (the desert nation has one of the most aggressive solar development programs in the world) and nuclear. With Grae's guidance the country signed an extensive agreement with the IAEA that allows for unannounced inspections and calls for the UAE to import nuclear fuel and send it elsewhere when it's spent. The UAE will never own nuclear fuel, just lease it. In so doing the UAE will become the first nuclear-powered country to voluntarily give up the enrichment of uranium and the reprocessing of plutonium, effectively eliminating its ability to make weapons.

It could also become the first country to build its commercial nuclear power industry around thorium—although for the moment UAE officials consider that a long-term solution. For now Lightbridge is helping the emirates build a uranium-based nuclear sector, which somewhat dims the luster of Lightbridge's thorium power ambitions.

The hiring of Jim Malone, the former vice president for nuclear fuels at Exelon, did not exactly confirm Lightbridge's thorium commitment, either. A veteran member of the nuclearati, Malone spent a decade at Exelon, the largest U.S. producer of nuclear power. When I spoke with him and Grae in August 2011, they made it clear that, while thorium fuel development remains a part of Lightbridge's plans, it is no longer necessarily the company's highest priority. In fact, Lightbridge was working on two other fuel elements, both based on uranium, one an all-metal variety. The company still hopes to make its thorium fuel technology commercially viable, but it will be mainly targeted at countries (like India) that lack extensive uranium reserves. "We don't choose among our children," said Grae—meaning that, as far as Lightbridge is concerned, the market will choose among thorium and uranium fuels.

In any case even Lightbridge's original thorium goals were too modest for thorium's true believers. Replacing uranium with thorium in conventional reactors, they say, is a half measure, like putting biofuel in a Hummer. What's needed is not just a new fuel but a new engine. The Lightbridge story is emblematic of the thorium movement as a whole: great technological promise dampened by the realities of today's nuclear market. Over and over I heard the same refrain: To carve out a place in the nuclear power industry, thorium must do more than offer the same performance with greater safety and fewer proliferation possibilities. It must be better. To see whether that can be true, it is necessary to compare new thorium reactors with the plans for future advanced reactors based on uranium.

TH90 • TH90 • TH90

THE SUMMER MEETING OF THE American Nuclear Society (ANS), held at a seaside resort in Florida in late June 2011, could have been a wake. It was three months after Fukushima. The nuclear power industry worldwide had lost the exuberance of the so-called nuclear renaissance of recent years, which saw plans for dozens of new reactors, costing billions of dollars and producing zero carbon emissions, to be installed worldwide. Nuclear power was being questioned and reassessed internationally, and three Western European countries— Germany, Switzerland, and Italy—had publicly announced that they would give up on it altogether and shut down their reactors. Two of the best-attended sessions at the ANS conference dealt not with new builds but with a lugubrious subject: lessons learned from Fukushima.

Shares in nuclear suppliers like Toshiba, owner of Westinghouse Electric Corporation and one of the major nuclear reactor vendors in the world, plummeted on the first day of the Fukushima disaster and had not recovered by the time of the conference. According to the antinuke website Nuclearbailout.org, the formerly resurgent industry had suffered eight defeats in state legislatures in 2011, including efforts to rescind nuclear moratoria in Wisconsin, Minnesota, and Kentucky—and no victories. On the surface it seemed a lousy time to be a nuclear power company.

Those clouds, however, masked a dawning technological revival in the nuclear power industry—one that could reshape the global power sector before 2022 as big nuclear suppliers like Toshiba and the French giant Areva, and major uranium miners like Cameco, bring long-awaited new reactor designs to market. Simply put, what happened at Fukushima may have slowed down the transition to new nuclear reactor technology, but it also made it more inevitable.

Fukushima, Kirk Sorensen told me, "marked the death of conventional light-water reactors." Evolutionary designs that utilities assumed would carry lower risks now faced regulatory hurdles raised by post-Fukushima fears, while more innovative reactors continued to move forward toward licensing and installation. Fukushima was like the devastating asteroid strike that, according to many paleontologists, doomed the dinosaurs and gave rise to the age of mammals.

To get a sense of where the nuclear power industry stands technologically, it's important to consider that the reactors at the Fukushima-Daiichi plant were designed in the 1960s. Fukushima No. 1, a "Mark 1" model, went critical in 1971, the year before a report by the Atomic Energy Commission recommended that the Mark 1 be discontinued for safety reasons. While other industries have been transformed by new technology since the early 1980s, innovation in nuclear power essentially ground to a halt after the 1979 Three Mile Island accident. Eighty-five percent of the electricity from nuclear plants today is generated by Gen II reactors like the Mark 1. It's as if nearly all the computers in use today were Altair 8800s, developed by Xerox in the early 1970s, except PCs don't blow up and spread radioactivity across the countryside.

By 2011 the design level had finally started to change. Long-overdue Generation III designs were finally starting to come online, and companies like TerraPower were developing fourth-generation reactors, to be built between 2020 and 2030. Some suppliers were even touting Gen-III+ machines, which was a little like putting new whitewalls on a 1999 pickup truck and calling it a 2011 model.

That summer the Tennessee Valley Authority signed an agreement with the nuclear contractor Babcock & Wilcox to build six small modular reactors, the first small modular reactors contracted for commercial power generation.

Ironically the site chosen for the small reactor project was Clinch River, the site of the ill-fated fast breeder reactor—a program that was outsized in every respect.

The government of France, meanwhile, announced plans to invest 900 million francs in nuclear power R&D, including thorium-powered systems.

The vast majority of Gen III and IV designs would run on traditional uranium fuel. Theoretically they'll be less expensive to build and more efficient on a cost-per-kilowatt basis than conventional reactors, and they will produce fewer toxic wastes to be stored indefinitely. Some designs, which I describe in more detail later in this chapter, are marvels of ingenuity; it's as if all the pent-up creative energy of reactor designers, stymied for decades, was released at one time. These designs could help bridge the way to the thorium-fueled future. "It's time for a new generation of nuclear power," declares a Westinghouse promotional brochure. Unfortunately, there's no guarantee that the new generation will ever make it to market.

The only Gen III reactor to actually be certified to date by the Nuclear Regulatory Commission (NRC) is Westinghouse's AP1000 machine. Based on a reactor certified in 1999 but never operated commercially, the AP1000 is the Ford minivan of reactors: simple, utilitarian, built for safety, and deeply conservative.[3] Nothing about the core design or the fuel cycle is innovative. It has what Westinghouse bills as a "passive core cooling system," which in theory will calm the reactor in the case of an accident, even if the operators do nothing. In theory, no electricity is required.

That all sounds pretty safe until you consider that the Fukushima plant had similar passive safety systems, and that, because the area had been so thoroughly contaminated, getting water to top off the cooling system became a major problem for weeks, not days. The AP1000 is basically a smaller, conventional light-water reactor with added passive safety systems. It's a minivan with extra side-door airbags and a roll bar.

TH90 • TH90 • TH90

LIKE OTHER SO-CALLED ADVANCED REACTORS planned by the industry, the AP1000 has hit bumps in the road to commercialization. Since Fukushima, environmental groups have petitioned the NRC to delay approval of the AP1000, a move that would put costly delays in the way of Southern Company and Scana

Corporation, the companies building the new multibillion-dollar reactors. The effect of Fukushima is likely to be paradoxical: it could help kill off half measures and somewhat updated reactor designs like the AP1000 because investors won't back them and the NRC, under the additional pressure of safety concerns, won't license them. To gain licensing for and build new reactors today, you need not only an acceptable but not too innovative design, but also plenty of patience, and very, very deep pockets. The nuclear renaissance is a matter less of science than of dollars and cents.

That said, some new reactor designs could shove uranium-fueled reactors into the late twentieth century, at least, if not the twenty-first. And thorium-based machines are, according to the global nuclear power cognoscenti, among them. The Generation IV International Forum, a collaboration of a dozen governments (including China, Russia, France, Japan, and the United States but not India) plus the European nuclear agency Euratom, has whittled the competing technologies to six (from more than 100), ranging from tweaks to existing systems to altogether new machines. Most are either advances to existing technologies or updated versions of older reactors. Only one is truly radical. For purposes of comparison, I will briefly examine each technology and its advantages and disadvantages.

The sodium-cooled fast reactor. The most familiar fourth-generation machine is, as you might guess, based on liquid metal fast breeders such as the ill-fated Clinch River project or the French Superphénix. Despite the LMFBR's checkered history, versions of sodium-cooled fast reactors are operating in Russia, South Korea, and India. According to Western analysts, China brought a prototype online in 2010.[4] (The Superphénix, a 1,200-megawatt sodium-cooled fast breeder was the target of one of the most audacious attempted acts of nuclear terrorism in history when five rocket-propelled grenades were fired across the Rhone River at the plant on January 18, 1982. The damage was negligible. The perpetrator of the attack is said to have been the notorious international terrorist Carlos the Jackal.)

Sodium, which is essentially transparent to neutrons and so does not slow down the fission reactions, has excellent heat-transfer properties and can be used at lower pressures than other coolants such as water and gas. Russia has been running sodium-cooled fast breeders since the 1980s. Proponents claim that producing energy from sodium-cooled reactors could be safer and less expensive than existing light-water designs (there's little proof of the first proposition and plenty of evidence against the second), and sodium-cooled reactors have the additional advantage of having actually been operated as commercial plants, not just as prototypes. "We really can build one," the nuclear scientist Robert Hill, of Argonne National Lab, told the *Economist*. But the dismal record of liquid metal fast breeders makes the sodium-cooled fast reactor, which is a type of liquid metal reactor, a puzzling choice for the new generation of reactor technology.[5]

The gas-cooled reactor (GCFR). The second design not cooled by water in the Gen IV lineup is the GCFR, which, like the sodium-cooled machine, would be a fast-spectrum reactor. Pressurized helium in a gaseous form would both cool the core and drive a turbine. Helium gas is noncorrosive and can be used at high temperatures. GCFRs would have one other notable advantage: they can use a variety of fuels, including thorium. The ability to run gas-cooled machines as breeder reactors could help make them more economical and less apt to increase nuclear proliferation than today's conventional systems; the increased efficiency means you need less fuel to produce a given amount of power.

No gas-cooled reactors have been operated on a commercial scale, though, and there is a historical irony to considering them for next-generation technology: the original design for the Hanford reactor, built to produce plutonium for the Manhattan Project, was a helium-cooled machine. Eugene Wigner called that high-temperature, helium-cooled machine "an engineering nightmare." His concept of a water-cooled reactor won out, and water-cooled reactors have been the dominant design ever since. Returning to gas-cooled machines today would be like deciding that steamships are the wave of the future for ocean voyages.[6]

Lead-cooled fast reactors (LCFRs). In their early nuclear submarines, the Soviets used reactors cooled with a blend of lead and bismuth, and LCFRs have gained renewed interest, particularly in Europe, since the 1990s. Using molten lead, or a lead-bismuth mix, could facilitate the creation of smaller modular reactors with a long-lived (some even say lifetime) core. In theory, such reactors could be supplied, prebuilt, to nonnuclear countries without giving them access to weapons-grade nuclear material.

In December 2009 a joint venture called AKME Engineering, one partner in which is the state-owned Russian nuclear company Rosatom, was formed to build thousand-megawatt lead-bismuth fast reactors based on the original design for Soviet subs. A prototype is planned for 2019. Like all the fast reactor designs, however, LCFRs are complicated to build and hard to run. There is no conclusive evidence that they can be produced and operated on a commercial scale.

The supercritical water-cooled reactor (SCWR) and the very high temperature reactor (VHTR). These are two of the other so-called Gen IV reactors, simply light-water reactors with new twists. Obviously reactor designers like to use superlatives (supercritical, very high) to distinguish the newer technologies from their more dated siblings; the actual guts of the machinery, though, are not that much different.

Supercritical refers to a fluid at such a high temperature and pressure— "beyond the critical"—that it exists in a state that is neither liquid nor gas (or steam, in the case of water). Because small changes in pressure or temperature result in large changes in density, supercritical fluids can be finely adjusted to exhibit desired properties. Supercritical carbon dioxide, for example, is used to decaffeinate coffee beans. In the case of a nuclear reactor, supercritical coolant water can be used to drive a turbine directly, eliminating the need for a secondary heat exchange system. That elevates thermal efficiency to the 45 percent range from the 33 percent typical of conventional pressurized water reactors. Simplifying the plumbing also lowers costs: the Generation IV International Forum has estimated that a reactor cooled with supercritical water could be built for $900 per kilowatt of capacity—about a quarter of the cost of advanced Gen III reactors like the AP1000.

The high pressure needed to maintain water in its supercritical state, though, requires a thicker pressure vessel. With relatively little water in the core, a reduction in pressure (caused by a pipe break, for instance) could lead very quickly to a catastrophic coolant loss. What's more, SCWRs will require the development of new alloys that resist cracking and corrosion—common problems in high-temperature reactors cooled with water, including the very high temperature reactor.

The VHTR is a bit of a gryphon, a mythical beast welded together from parts of the eagle and lion. Like the original X–10 pile at Oak Ridge, the VHTR would use graphite, rather than water, as the moderator and helium gas (as in the gas-cooled fast reactor) as the coolant. Other Gen IV designs attempt to close the fuel cycle through various forms of reprocessing and reuse of spent fuel; the VHTR is a once-through cycle that would do little to reduce the waste or the threat of proliferation posed by current light-water reactors. Nevertheless, in 2009 the Obama administration announced $40 million in R&D funding for the Next Generation Nuclear Plant project, which is based on the VHTR. *Next generation,* in nuclear power, is a fungible term. As of early 2012, thorium-based reactors were not part of the federal government's next-gen nuclear R & D program.

All these are new and ingenious ways of solving old problems—the inherent problems of uranium-based solid fuel reactors. Nuclear engineers have spent a half century building ever more sophisticated systems that will essentially prevent uranium from doing what it wants to do—to open the floodgates a crack without washing away the dam, as it were. Proponents of LFTRs want to build machines that allow thorium to do what it wants to do. And, in fact, the molten salt reactor—championed by Alvin Weinberg, developed at Oak Ridge, and abandoned on the eve of the first energy crisis—is the final Gen IV design on the forum's shortlist of competing designs.

TH90 • TH90 • TH90

IN MAY 2011, AFTER I ATTENDED the third Thorium Energy Alliance conference in Washington, D.C., I dropped in on a meeting of the vaunted Blue Ribbon Commission on America's Nuclear Future, known as the BRC, that was held in the basement meeting room of a downtown Washington hotel. The occasion was an update on the NRC's responses to Fukushima and a review of the draft report from the Reactor and Fuel Cycle Technology Subcommittee, which was considering new approaches to the challenges of next-gen reactor design and waste disposal. Bespeaking the disfavor in which the nuclear industry still finds itself, renaissance or no, the meeting drew a few dozen spectators—some interested industry officials, some Hill staffers, a few reporters, and that was about it.

The BRC is one of those full-frontal Washington offensives that get announced with great fanfare and then tend to peter out in endless meetings, voluminous reports that no one but wonks ever reads, and few effective policy changes. It's like a religious conversion: the mere fact of forming the commission, it was hoped, would produce a magical solution to the problem (specifically, in this case, the problem of nuclear waste; more generally, as the commission's full name made clear, the future of nuclear power in the United

States). In reality, the Blue Ribbon Commission was an attempt to reconcile two fundamentally opposed positions: the Obama administration's strong support for nuclear power, in theory, as a primary source of carbon-free energy on the one hand, and its adamant opposition to the Yucca Mountain storage facility for nuclear waste on the other. By this time it was clear that Yucca Mountain would never be built. Without a place to store the spent fuel and toxic waste from conventional nuclear plants, the industry could not move forward. Appointed by President Obama soon after he took office, the BRC was supposed to be the sword that severed this Gordian knot. Although it was headed by the beltway eminences Lee Hamilton (former chair of the House Foreign Affairs Committee) and Brent Scowcroft, the former U.S. national security adviser, its edge was dull. By the summer of 2011 little had come of the BRC beyond a series of draft reports, and it looked as if no firm conclusions, much less a plan, would emerge before the end of Obama's first term. The Blue Ribbon Commission was on its way to proving the old adage that the best way to kill an idea is to assign a committee to study it.

The meeting I attended took place almost exactly two months after the tsunami broke over the Fukushima-Daiichi plant. The session quickly devolved into a Laurel-and-Hardy skit, as an NRC bureaucrat named Lawrence Kokajko was grilled by the commissioners, particularly Hamilton, a man used to having bureaucrats tremble before him. Kokajko said that, at this time, the NRC had no information that "would cause us to doubt the safety of the current operating [reactor] fleet." Hamilton wanted more specifics. Kokajko restated without answering. Much posturing ensued. It was like a reenactment of the entire history of nuclear power in the United States: plenty of hot air, a certain amount of power produced, and little real progress.

The only one whose performance came off well during the meeting was Per Peterson, a professor of nuclear engineering at the University of California at Berkeley.

I'd met Peterson at the Thorium Energy conference earlier that week. He is slender, with a high forehead, thinning dark hair, and piercing brown-black eyes behind rimless glasses. The lips of his thin mouth curl slightly upward at the sides, giving him the suggestion of a perpetual smirk. Brilliant and direct, he has the air of a man with a low tolerance for stupidity. Peterson chairs the Reactor & Fuel Cycle Technology Subcommittee of the BRC. He and his team of grad students have carried out several experiments related to advanced reactor design and thorium power, and Peterson has become something of a hero to the thorium movement. His work focuses on what's known as the pebble bed advanced high temperature reactor (PB-AHTR), a machine that could provide a bridge to the LFTRs of tomorrow.

Based on two older technologies—coated particle fuel (pebbles) and coolants made from molten salt—PB-AHTRs offer the advantages of both molten salts and modular design that make them easier to manufacture, license, and site. The pebble-based fuel reaches full depletion (that is, it gives up all its available energy) in less than a year, meaning that demonstration reactors could test different fuel types much more rapidly than other designs. And PB-AHTRs, in theory, would use far less natural uranium and leave behind less spent fuel than today's designs. As in LFTRs, the molten salt coolant is chemically inert and contained at low pressure, making Peterson's design safer than light-water reactors.

"The economics are much better than current reactor technologies," Peterson told me. Proving that high-temperature pebble bed machines would successfully produce commercial power would pave the way for thorium-based reactors. The link seemed somewhat tenuous to me, but having a mainstream physicist of Peterson's stature—and his access to research funding—in its camp was a major coup for the thorium movement.

Peterson, however, is a realist. He had good news and bad news for the thorium fans in the audience: The bad news is that reactors that use new materials or new fuels will require multiple decades, at least in the United States, to be funded and licensed. The earliest that a commercial thorium-based reactor could come online, in Peterson's view, would be 2032. That's not soon enough to solve the energy crisis. And it's not soon enough for thorium advocates like Kirk Sorensen, who wants to see a LFTR built before 2020.

TH90 • TH90 • TH90

BESIDES LENDING THE BRC PROCEEDINGS an air of rigor and incisiveness they otherwise lacked, Peterson's main task was to present the draft report from his subcommittee. To be included in the Blue Ribbon Commission's final report, which was delivered to the president in January 2012, the Peterson report is as close as the United States is likely to get to mapping out a path forward for new reactor technology; while it was couched in the cautious idiom of official Washington, the draft echoed loudly the call for a second era of nuclear innovation. It even mentioned thorium several times. Specifically, the subcommittee called for a solution to the nuclear waste problem through a return to uranium recycling schemes, long abandoned in this country but a key element of nuclear power industries in France, India, and even Japan.

The events at Fukushima "underscored the importance of treating spent fuel management and storage as a central part of the safety regime," Peterson and his team concluded. "Technological advances hold promise for improving the safety of nuclear energy systems—ensuring that this promise is realized must be a priority of U.S. nuclear policy, with respect to both RD&D [research, development and deployment] investments and deployment decisions."[7]

The draft report called for "a major international effort, encompassing international organizations, regulators, vendors, operators, and technical support organizations," to develop and operate safe nuclear power plants and safely manage nuclear wastes worldwide. Like Jim Kennedy and John Kutsch, Peterson and his colleagues also raised the specter of diminished U.S. competitiveness: "It is in our nation's interest to retain [our] leadership role" in nuclear technology, the draft states, ignoring U.S. cession of that role back in the 1980s. "Regardless of one's view of the nuclear industry's near- and longer-term prospects more generally . . . there are countries planning to increase their nuclear energy investments—in some cases substantially—while other countries that currently lack nuclear energy infrastructure are interested in developing it."[8] Giving up on nuclear power would mean not only remaining dependent on fossil fuels but giving away any influence over the shape and spread of nuclear technology—both power reactors and weapons production.

There followed the predictable call for a "sustained, strategically targeted, and well-coordinated federal RD&D effort" to bring to market "game-changing" technologies that would accomplish the nation's energy security as well as its economic and safety goals. Among those game changers, according to the subcommittee, is thorium: a thermal spectrum, high-temperature molten salt reactor using thorium could solve many problems associated with current "once-through" (lacking fuel recycling capabilities) uranium-based reactors.

And then came the usual caveat: "Such systems could potentially offer many of the combined benefits of the alternatives listed. However, these systems have not received systematic study and the component technologies for these types of systems are less well developed."[9]

To anyone familiar with the Molten Salt Reactor Experiment at Oak Ridge in the 1960s, the suggestion that "these systems have not received systematic study" is laughable. Many millions were spent, thousands of hours were put in, and reams of studies and results were produced in the systematic study of thorium-based MSRs. To be sure, much more has been carried out on fast breeders, which remain near the top of the lists of Gen IV reactors (and which are cited in the Peterson report as one of three "representative advanced nuclear energy systems," along with once-through high-temperature reactors and modified light-water machines—a discouraging representative sampling). It could be argued that the net result of the fast breeder programs has been much less encouraging than the study of MSRs.

Peterson himself is familiar with this history. His committee's report is a valiant effort to shift the discussion of nuclear power away from fear and rhetoric and toward the future. The cautious language of official scientese, though, meant that thorium could receive no more than an obligatory mention in the report of his subcommittee. The most important recommendation from the reactor and fuel cycle report was to reshape the NRC to make it capable of operating effectively in a Gen IV world: "We concluded that a portion of the federal government's nuclear-energy R&D investment should go to the NRC and be directed for work to develop a regulatory framework to perform supporting, anticipatory research for advanced reactors and fuel cycle technologies significantly different than those we use currently," Peterson explained to me in an email. The portion he had in mind was not large: 5 to 10 percent. The report does not spell out what, exactly, would constitute "anticipatory research."

Clearly, the federal technological bureaucracy has not progressed far from the days of Milton Shaw and WASH-1222. In the reactor and fuel cycle technology report, thorium was basically an afterthought, as it has been since the beginning of the Atomic Age. Until a new generation of nuclear technologists and executives rises to places of authority in the industry, there's little prospect of that changing.

Take a step back, though—and get out of Washington—and the picture resolves into a wider pattern. Not only are many countries and companies aiming to solve the energy crisis using thorium as the basis for the twenty-first-century nuclear power industry, but the movement toward liquid-fueled thorium reactors forms the next, and perhaps the ultimate, advance in the centuries-long quest for safe, sustainable, and efficient forms of energy. This search has two parallel streams: the progress from solid to liquid forms of fuel, and the elimination of waste products.

TH90 • TH90 • TH90

THE BIRTH OF THE MODERN ERA OF HYDROCARBONS is usually traced to 1859, when Colonel Edwin Drake drilled the first oil well in North America, at Titusville, Pennsylvania, or a bit further, to 1848, when the first modern well was drilled near Baku, Azerbaijan, on the Caspian Sea by the Russian engineer F. N. Semneyov. In fact, though, humans' fascination with, and search for, "the magic liquid" goes back more or less to the dawn of recorded history. Bitumen, the viscous form of petroleum, was tapped in Mesopotamia as far back as 3000 B.C. In the *Iliad* the Trojans poured "unwearied fire"—a weaponized form of liquid petroleum—on the Greeks' ships, causing "a flame that might not be quenched." The gas seeps found commonly across the Middle East fueled permanent flames, helping to create the fire-worshiping cults that Moses abominated. Used by the Byzantines in the countless wars of the Dark Ages, *oleum incendiarum* (Greek fire) was an early form of the ultimate destructive weapon, the atom bomb of its time.[10]

A simple insight underlay the fascination: liquid fuel is more efficient, more transportable, easier to build into machines, and more dense in terms of energy per unit of volume than solid fuel. From the time people first began systematically using fire for heat, illumination, and cooking, 300 or 400 millennia ago, they have sought ways to make it easier to use. It did not take great ingenuity to realize that liquid forms of flammable fuel were better than solid ones, whether you were carrying them across the desert or heaving them at enemy ships. From burning branches to charcoal to coal to petroleum, the march of civilization is also the progressive compaction and liquefaction of primary fuels.[11]

With the discovery of the boundless energy at the heart of the atom, that progress seemed to have reversed. Solid enriched pellets of uranium were the most energy-dense forms of matter ever known. What need is there to use liquid fuels (petroleum) when you have such a marvelously compact (and relatively lightweight) source of energy as the uranium atom?

Plenty, as it turns out. In chapter 3, I listed the advantages of liquid nuclear fuels over solid ones. Most of those advantages apply to fossil fuels as well. You can't fly a jet with coal. You can launch a rocket using solid fuel; the defunct space shuttle program used solid fuel to boost the rocket into orbit—a stage that requires huge amounts of thrust, rather than efficiency, and no throttle capability—and liquid fuel once in orbit, to enable more efficient and flexible power. Once you start up a solid fuel booster, you can't shut it down: it continues to burn until the fuel is consumed. That's a familiar scenario to anyone who has ever been faced with cooling off a runaway solid-core uranium reactor.

Whether the fossil fuel is solid or liquid, though, you still have the problem of waste. Coal plants leave behind coal ash (which is more radioactive, ounce for ounce, than nuclear plant waste), and huge amounts of carbon are emitted into the atmosphere (a typical coal plant generates 3.7 million tons of CO_2 a year, according to the Union of Concerned Scientists); liquid petroleum, natural gas, and gasoline are only slightly cleaner. The same goes for solid-core uranium reactors: as the Yucca Mountain debacle demonstrates, the issue of nuclear waste is the single biggest obstacle to the renaissance of nuclear power in the United States, from a perspective concerned about waste storage and the proliferation of nuclear weapons. Liquid fuel reactors, specifically LFTRs, show the greatest promise for overcoming the waste dilemma. Thorium advocates, in fact, say that LFTRs will essentially eliminate the nuclear waste problem altogether.

As you might guess, the situation is not quite so simple. Understanding it fully requires reviewing the origins and makeup of nuclear waste. To do so I return to a familiar figure in this story: Eugene Wigner.

TH90 • TH90 • TH90

THE MOST PROBLEMATIC LONG-TERM NUCLEAR WASTE is the transuranic elements. Created in the reactor's nuclear forge, the core, transuranics are simply elements beyond uranium—their atomic numbers are higher than that of uranium, number 92. None of these elements occurs in nature in more than trace amounts. The one that is most important in controlling the spread of nuclear weapons is plutonium-239. Often referred to as the deadliest substance on Earth, it is no more toxic than many substances routinely produced by industry every year, including the by-products of radon gases. With a half-life of only 24,200 years, though, plutonium is intensively radioactive. About ten kilograms of nearly pure PU-239 are required for a single nuclear weapon. The spent fuel from nuclear reactors is contaminated with other, lesser isotopes of plutonium (PU-240, PU-241, and so on), which are difficult to separate from the desirable plutonium-239. Still, with its potential for bomb making and its capacity to poison humans when inhaled at low doses, plutonium lies like a dark fruit at the end of a series of transmutations undergone by uranium.[12]

A thousand-megawatt light-water reactor produces about 290 kilograms of plutonium a year. Total world production of reactor-grade plutonium is about 77 tons a year; approximately 1,510 tons have been produced to date. You can buy it from Oak Ridge if you have the right qualifications and clearance, which in practice eliminates nearly everyone. ("Volume discounts available," the ORNL website exclaims.) Since we started creating it in the early 1940s, plutonium has frustrated every attempt to dispose of, cease production of, and wish it away. It is the distilled essence of militarized nuclear power, and its malign influence can be felt behind every decision made in the industry since the 1950s.

The other type of radioactive waste, fission products, was first understood by Eugene Wigner, who explained the early failure of the X-10 reactor at Oak Ridge under the Manhattan Project by predicting that the fission reaction was being poisoned by the buildup of xenon-135, which has a high capacity to absorb neutrons. As xenon accumulated in the reactor, the rate of fission reactions slowed, and eventually the chain reaction fizzled out. This problem was solved in the X-10 by simply adding more fuel rods. However, the buildup of xenon and other contaminants eventually ruins conventional solid uranium fuel rods; that's why they must be replaced after only a small percentage of the uranium is consumed (less than 1 percent of the uranium that is mined is fissioned in today's reactors). Uranium reactors are self-defeating: the fission products eventually contaminate the fuel so that it must be removed and replaced, requiring the reactor to be shut down and creating spent fuel rods that must be safely stored for many millennia.

The real threat of fission products is not in the reactor, of course: it's in the environment, where high-level wastes remain radioactive for thousands, in some cases millions, of years; these must be shielded to prevent radiation from poisoning the environment. The amount of high-level waste worldwide increases by about 13,230 tons a year; that number will rise sharply in coming decades if predictions for new reactors prove close to being accurate.

As I've noted, several countries, including France, Japan, and Russia, now have active reprocessing programs to remove the fission products and convert spent fuel to reusable fuel elements. Reprocessing converts the waste into a form appropriate for permanent geologic disposal while recycling the useful unburned fuel back into reactors. Another significant conclusion of Peterson's Reactor and Fuel Cycle Technology Subcommittee was that the reprocessing of spent fuel—abandoned in the United States under President Gerald Ford in the 1970s—should be a part of any future nuclear power scenario. Reprocessing nuclear waste would dramatically reduce the amount of new waste added to our already huge pile of radioactive waste (most of it stored, as at Fukushima-Daiichi, in spent fuel pools at nuclear power plants). One problem with this scenario is that, because uranium has been so inexpensive (and, as a percentage of operating costs, such a minuscule expense for nuclear power producers), there's little market incentive to recycle spent fuel. Since no politically acceptable solution for storing spent fuel is on the horizon, though, reprocessing is an absolute requirement for a new nuclear era in the United States.

Most Gen IV reactors are billed as closing the fuel cycle, that is, they include some form of fuel recycling. Fast breeders are considered especially useful for this because, with their overabundance of neutrons, they can consume less-than-ideal fissile materials. Reprocessing, though, comes with its own set of disadvantages, one of which is that it's like a pyramid scheme: it requires more and more infusions of cash (in the form of energy) to produce diminishing returns.

"It might take three reactors to create enough fuel to be recycled to fuel one reactor, and less and less each generation," Sorensen explains on the *Energy from Thorium* blog. "It's a losing game (which is why the nuclear industry is always trying to tell you about the brave new world of fast-spectrum reactors)."[13]

Thorium advocates argue that thorium essentially solves the problems of waste storage and proliferation because the total volume of spent fuel is smaller (thanks to thorium's more favorable burn-up profile), the proportion of high-level wastes requiring long-term storage is smaller, and the length of time they must be sequestered is shorter as well. However, thorium-based solid fuel reactors do produce plutonium. And their remainders are actually even more radioactive than the spent fuel from a conventional reactor. Paradoxically, that's a good thing: it makes handling the wastes, and building bombs from them, more hazardous than doing so with the waste from conventional reactors. This is the subject of hot dispute. To summarize the arguments, I'll step back and review the most common objections to thorium power.

TH90 • TH90 • TH90

IN REPORTING ON THE THORIUM POWER MOVEMENT, I heard plenty of reasons why it would never work. After a year or so I classified them into three categories: market barriers, challenges related to waste and proliferation, and what I came to call the traditionalist argument.

The market-based argument is simple: the nuclear power industry has a fuel today that is abundant and inexpensive. Why should it switch to a new, relatively unproven fuel? These assumptions are faulty (uranium may well not be inexpensive and plentiful much longer—see the comments of Srikumar Banerjee, chair of India's Atomic Energy Commission, from chapter 7). More important, this argument does not take into account the broader costs and risks of uranium-based nuclear power, which have been highlighted by the Fukushima-Daiichi accident. There's little chance of nuclear power's fulfilling its promise until those costs are driven down—by shifting to thorium power.

The waste and proliferation issues are more complicated, and I will break them down into four elements.[14] In distilled form they sum up the objections to thorium from both the nuclear establishment and antinuclear groups.

1. *The use of enriched uranium or plutonium in thorium fuel to ignite the fission reaction carries proliferation risks, and U-233 is as useful as Pu-239 for making nuclear bombs.*

    This is the central claim of those who dismiss thorium's prospects for reducing the nuclear waste stream: Solid-fuel thorium reactors produce both U-233 (the fissile daughter element of Th-232) and plutonium, so what's the difference? What's more, thorium reactors require low-enriched uranium or plutonium to initiate the fission reaction, thus creating more material that can be refined into bombs.

The kernel of truth here is that the U-233 (and thus the plutonium as well) created in the transmutation of thorium is contaminated by U-232, one of the nastiest isotopes in the universe. With a half-life of less than 70 years, U-232 decays into the radioisotopes bismuth-212 and thallium-208, which emit intense gamma rays that make it very, very hard to handle and transport (not to mention reprocess) and that would very likely destroy the electronics of any weapon into which they were built. Theoretically, it's possible to make a bomb with U-233, but plutonium is much easier to make and does not come with the problematic U-232. Militaries will always opt for plutonium and U-235, because they can't afford to expose their personnel to the deadly risks of U-232. As for terrorists, they'd be better off simply buying natural uranium on the open market and finding a way to enrich it. The United States reportedly tested bombs with U-233 cores in the late 1950s, but no country has ever included it as a material as a part of its nuclear weapons program. It's useless even for the most zealous of hypothetical suicide bombers, because they'd probably never reach their target.

2. *Most proposed thorium reactors require reprocessing to separate out the U-233 for use in fresh fuel. As with conventional uranium power plants that include reprocessing, bomb-making material is separated out, making it vulnerable to theft or diversion.*

This is a tired canard. Never mind that every nuclear fuel cycle currently in production or contemplated generates "bomb-making material"—this statement ignores the realities of weapons building. Most Gen IV designs described in this chapter involve fuel recycling; indeed, as the Peterson report stated, recycling is critical to the future of nuclear power. To be sure, reprocessing spent fuel rods from a solid fuel thorium reactor is not a simple matter, whether you're making bombs or new fuel. But it's important to note that, as with all these arguments, external reprocessing is necessary only for solid fuel reactors—not LFTRs. Alone among advanced reactor designs, LFTRs have the capacity to reprocess the fuel *in the reactor building itself,* while the reactor is operating. There's no opportunity for diversion unless you raid the entire plant, shut down the reactor, and figure out a way to separate and abscond with the weaponizable isotopes. Good luck with that.

3. *The claim that radioactive waste from thorium reactors creates waste that would have to be isolated from the environment for only 500 years, whereas irradiated uranium-only fuel remains dangerous for hundreds of thousands of years, is false. Thorium-based reactors create long-lived fission products like technetium-99 (its half-life is more than 200,000 years), and thorium-232 is extremely long lived (its half-life is 14 billion years).*

This argument ignores the larger context. The volume of fission products from thorium-based solid fuel reactors is about a tenth of that from conventional reactors. What's more, in small amounts, many of these fission products have become common in modern life. Technetium-99, for example, is powerful stuff, worthy of respectful treatment; it's also commonly used, in a slightly altered form, in medical imaging procedures. Millions of patients ingest it every day without significant risk. The amounts of technetium-99 produced in solid-fuel thorium reactors would be negligible; in LFTRs it would be processed off along with other fission products and largely recycled. Some geological storage will be required, but in general waste from LFTRs decays to safe, stable states within a few hundred years—far less than the millennia required for the by-products of uranium reactors. As for Th-232, it's long lived but safe. The longer-lived a radioactive element is, the lower its radioactivity—with its very long half-life, Th-232 is an exceedingly weak producer of radiation. It is so common that it's found in small amounts in virtually all rock, soil, and water. You could sleep with it under your pillow and suffer no ill effects.

4. *Reprocessing of thorium fuel cycles has not been successful because ura-nium-232 is created along with uranium-233. U-232, which has a half-life of about 70 years, is extremely radioactive and is therefore quite dangerous in small quantities.*

U-232 is indeed extremely radioactive, but its brief half-life means that in less than a century half of it will have decayed to a stable form. Because isotopes decay at a geometric rate (50 percent of half of the original material, or one-quarter of the original, is still radio-active after another 70 years, then one-eighth, one-sixteenth, and so on), the decrease in radioactivity drops off quickly. Many, many hazardous materials are put in storage for centuries. We do not object to them.

To summarize, the most common objections to thorium power from the perspective of radioactive waste and the proliferation of nuclear weapons are inflated for solid fuel reactors, and they simply do not apply to LFTRs. That leaves the traditionalist argument, which essentially echoes Milton Shaw and the WASH-1222 report from 1972: *It can't be done because it has never been done before.* When I heard this brand of defeatism, it always came from someone with a vested interest in the current nuclear power establishment. I'll explore the traditionalist argument in more detail in the final pages of this book. By early 2012, even as the Blue Ribbon Commission presented its blueprint for a nuclear future, the naysaying by a growing chorus of thorium advocacy, as the early believers—Kirk Sorensen, John Kutsch, Robert Hargraves, and others—were joined by a new group of converts, some of whom were businessmen with more concrete plans, and more access to capital, than the original enthusiasts. The thorium revival was entering a new phase.

NINE

THE BUSINESS CRUSADE

I n late January 2011, I was at a small gathering of thorium activists at Oak Ridge when Kirk Sorensen stood up and made a startling announcement: he was leaving his job at Teledyne Brown (where he'd been the chief nuclear technologist since 2010) to start his own company, Flibe Energy.

This was surprising not just because Sorensen already had what he'd described to me as a dream job (and one that would, he said, lead to the building of liquid fuel thorium reactors [LFTRs]). Named for the liquid fluoride solution in which thorium fuel is dissolved in a LFTR, Flibe was the first U.S. company formed expressly to develop, build, and run thorium-based liquid-core reactors. It was also a huge risk. After his presentation, I pulled Sorensen aside to ask exactly how he hoped to pull it off.

"I can't go into details right now," he told me. "But in a few weeks we'll have some very, very exciting news to share."

Specifically, Sorensen expected a significant infusion of funding from wealthy investors in Silicon Valley. This, he said, was how the thorium renaissance was going to be created: not through federal grants or national laboratory research programs but by a good old-fashioned start-up backed by the greatest entrepreneurial engine in history, the high-tech venture funds of Northern California.

Sorensen told me he was pursuing funding from Silicon Valley. A few weeks went by, and then a few more, and Sorensen—who'd brought in a partner, Kirk Dorius, a mechanical engineer and intellectual property lawyer, and occupied a modest set of offices in Huntsville, not far from the Marshall Space Flight Center—was still awaiting word from his deep-pocketed contacts in Silicon Valley. He may have had a world-changing energy technology, but he was now in the position of hundreds of small tech start-ups that crop up like mushrooms every year: armed with a great idea, a burning passion, and a business plan but not enough hard cash to bring it to fruition.

In that sense, Sorensen's brainchild exemplified the thorium movement as a whole: a powerful idea that seemed incontrovertible when explained simply but that faced large and seemingly immovable hurdles, funding not the least of them. In a series of stories on thorium power that ran in July 2011 in the *Nashville Tennessean,* Paul Genoa, director of policy development at the Nuclear Energy Institute, summed up the nuclear establishment's opposition succinctly: "There's a huge investment and infrastructure [in conventional nuclear power] in this country that goes back 50 years. You don't just walk away from that and try the shiny new toy, even if the shiny new toy might work better."[1]

Flibe's shiny new toy was a liquid fluoride thorium reactor, the twenty-first-century version of the thorium-based molten salt reactor (MSR) imagined by Alvin Weinberg nearly 50 years earlier. Flibe's reactor would have a classic seed-and-blanket design, with an outer layer of fertile thorium tetra-fluoride in a solution of lithium and beryllium (^{7}LiF-BeF$_2$, the flibe of the company name) surrounding an inner core of uranium-233 in the same lithium-beryllium cocktail, bred from the Th-232 of the blanket. The blanket salt and the fuel salt would be continuously recycled to breed new fuel and to burn off poisonous waste products. In a primary heat exchanger, the hot salt from the core would heat coolant salt, which in turn would heat gas in the secondary exchanger to drive a turbine. In a throwback to Alvin Weinberg's massive Middle Eastern facilities, waste heat from the gas turbine would be used to desalinate seawater. An external cooling fan would keep the freeze plug frozen solid at the bottom of the core. In the event of total power loss, the plug would melt and the radioactive fuel would drain into a sealed underground tank. Inexpensive power and freshwater from an inexhaustible fuel source, produced in a machine impervious to earthquakes, tsunamis, and terrorists, with minuscule waste and essentially zero risk of contributing to the proliferation of nuclear weapons. The ultimate power source. All Flibe Energy needed was a deep-pocketed investor.

Unlike most owners of companies working on thorium reactors, Sorensen didn't plan to just develop the technology and license it; he wanted to build power plants. That, of course, is unlikely to happen in the United States under current market and regulatory conditions. Sorensen, though, had a couple of ideas for getting around those problems, one of which was the military. For licensing purposes, he hoped to go through DARPA—the Defense Advanced Research Projects Agency, the source of many high-tech advances, including what became today's Internet. Colonel Paul Roege, a program manager in DARPA's Strategic Technology Office, is effectively the head of the Pentagon's advanced energy concepts programs, and, like the Berkeley physicist Per Peterson, Roege represents a great hope to the thorium movement. The U.S. military operates at least 100 nuclear reactors around the world, and it is spending billions to develop small, distributed sources of power that are not tied to the electrical grid—like, say, self-contained thorium-powered reactors.[2] Sorensen's hope is that the Pentagon may see its way to building a prototype LFTR—which would not have to go through licensing and approval by the Nuclear Regulatory Commission (NRC).

When I spoke with him, Roege declined to speculate on how likely that scenario was. The Pentagon is basically considering whether to examine the question of whether it should explore thorium reactors. That's not a business plan. Flibe Energy has another good idea, though, and it has nothing to do with producing energy.

TH90 • TH90 • TH90

AFTER THE FUKUSHIMA-DAIICHI ACCIDENT, there was a brief run on supplies of iodine-131. An isotope of iodine produced in specialized reactors, iodine-131 is used to prevent thyroid cancer from radiation exposure. Many people on the West Coast of the United States and elsewhere feared that widespread fallout from the reactor explosions at Fukushima could lead to millions of cases of exposure and to iodine-131 shortages. Those fears proved unfounded, but the alarm highlighted a long-festering crisis that many governments and medical professionals are just starting to understand: a shortage of what are known as medical radioisotopes.

Nuclear medicine has been around since the days of Marie Curie, but since the early 1990s it has become a multibillion-dollar industry. Cobalt-60, for instance, is widely used in radiation therapy for cancer. Technetium-99 is a tracer element used in medical imaging. Strontium-89 is used to relieve the pain associated with bone cancer, and cesium-131 has shown great promise in treating difficult cases of prostate and brain cancer. None of these substances is found in nature; they must be produced in nuclear reactors. Those reactors are aging quickly, and they're not being replaced.

The Chalk River reactor in eastern Ontario supplies about 90 percent of the U.S. supply of medical isotopes. Chalk River is 54 years old, and it is showing its age: it has been shut down twice in the last few years, including a yearlong closure in 2009–2010 that caused a worldwide isotope shortage and price spike. U.S. nuclear medicine providers were forced to turn to overseas suppliers to keep the clinics running. Chalk River was supposed to be permanently shuttered in 2005 and replaced by two reactors built by Atomic Energy Canada Ltd., but that plan was abandoned due to licensing issues and amid intense public opposition to the building of the new reactors. Chalk River is scheduled to close for good in 2016.

Almost all of the world's supply of molybdenum–99, the most widely used medical isotope, comes from only five research reactors in Canada, France, South Africa, the Netherlands, and Belgium. Although some have switched to using low-enriched uranium, most still use weapons-grade material, and all are more than 40 years old. Various bodies in the last few years, including the International Atomic Energy Agency, have warned that new sources, using low-enriched uranium not suitable for weapons, must be developed or the millions of people who are examined and treated annually using radioisotopes could be out of luck. Several companies have sprung up to find new ways, using advanced technologies, including small aqueous homogenous reactors, to develop the rare isotopes. Those projects are not expected to produce results for at least five years, most likely a decade. Cancer-fighting groups have called on the U.S. government to recreate a domestic medical isotope industry—an unlikely prospect in the post-Fukushima era of increased nuclear fears and drastic budget cuts.

The only reactors that produce medical isotopes are shutting down, and what has so far been a shortage could quickly turn into a crisis. Therein lies a business opportunity, believes Kirk Sorensen.

LFTRs, it turns out, are particularly good at producing medical isotopes, without using highly enriched uranium. Sorensen predicts that small, non–power-producing LFTRs could make half a billion dollars a year from medical isotopes alone. The medical isotope market is Flibe Energy's short-term path to profitability while building power plants. This was the barb in the hook of Flibe's business plan: on its way to making money in the $2 trillion or so worldwide energy market, Flibe would quickly become a major supplier in the multibillion-dollar market for medical isotopes.

Sorensen, of course, is not the only technologist to see the potential of the medical isotope market. In January 2011 the French nuclear giant Areva said that its radioisotope unit, Areva Med, had won approval from the FDA to begin clinical trials in the United States of cancer treatments based on lead-212. Areva has developed a proprietary process to separate lead-212 from thorium, which the French company produces in large quantities in its mining and reactor operations. Lead-212, whose chemical symbol is Pb-212, is a short-lived isotope with a half-life of ten hours; it decays to bismuth-212. In animal tests, Pb-212 has shown the potential to destroy cancer cells while leaving unharmed the body's healthy, normal cells. (Briefly, Pb-212 can bind to a monoclonal antibody—the immune system's attack cell. The combined antibody-radioisotope then attacks tumor cells. The treatment is known as alpha radioimmunotherapy.) Phase I clinical trials, expected to take two years, began in mid-2011. If successful, lead-212 radioimmunotherapy could be effective against some of the deadliest forms of cancer, including ovarian, pancreatic, colon, and breast cancers.

Foreseeing the medical isotope shortage, Areva is building an isotope separation facility in the Limousin region of southwest France. At the same time U.S. companies are trying to elbow their way into the medical isotope market: in 2011 a little-known company called Advanced Medical Isotope Corporation (AMIC), based in Kennewick, Washington, claimed that it was the leading domestic producer of molybdenum-99, used to produce the valuable isotope technetium-99m, which is used in more than 16 million imaging procedures a year. Headed by James Katzaroff, a former Wall Street analyst, and Robert Schenter, an expert on neutron decay and isotope production, AMIC uses a proprietary linear proton accelerator technology to create medical isotopes. It is a publicly traded company with a market value of nearly $20 million.

Flibe Energy, it goes without saying, is not yet so well funded. The medical isotope shortage added another layer of irony to the thorium revival: highly valuable, medically critical substances were growing scarce partly because of concerns about the proliferation of highly enriched uranium, while LFTRs, which could provide a clean, safe source of those isotopes in addition to supplying a large fraction of the world's growing energy demand, went begging. In the strange fashion of the nuclear industries, this seemed predictable, even normal. That normality was what people like Kirk Sorensen had dedicated themselves to shaking up.

In the meantime, Flibe survived on investments from family and friends—"enough to keep the lights on." Sorensen and Dorius were not drawing salaries. Sorensen's wife, now with four young children at home, was totally supportive, he said: "She was even more on board than I was" when he quit his job at Teledyne Brown. And Sorensen himself spoke in the terms of a former missionary.

"If I didn't do this, if I stayed in my safe job, twenty years from now when the world comes to be what I see coming, how much would I hate myself?" he asked me. "I knew about something that had the potential to change the world—and that this outcome was not destiny.

"At least this way I can say I went after it. My kids will know that I wasn't so afraid that I didn't do it."

TH90 • TH90 • TH90

FLIBE AND LIGHTBRIDGE ARE THE BEST-KNOWN companies created to develop thorium power technology, but they are hardly the only ones. By August 2011, more than two years into my exploration of the nuclear future, it seemed as if every month brought word of another far-flung, far-fetched thorium power start-up. Some were in the United States; some were in Europe. At least one was in Japan, which has essentially no uranium reserves. Some were connected to major vendors and engineering firms; some were independent. Few were well funded.

I began to realize that thorium reactor companies were like garage bands: it didn't take much to start one, the equipment was freely available, and, like grunge rock, assembling a new reactor didn't really seem all that hard. That was when I met Hector Dauvergne.

I heard of Dauvergne's company, DBI (for Dauvergne Brothers International), through a chance mention in an email. I had trouble figuring out exactly what kind of machine the "DBI thorium reactor" really was. The fuel apparently was solid, but Dauvergne claimed that it was not a conventional machine adapted from uranium fuel. It had been designed specifically to burn thorium. A breeder, it would continually consume the U-233 created in the core. Waste would be reduced by 90 percent. The small amount of spent fuel left over at the plant's decommissioning would be stored underground. The cost per kilowatt would be less than seven cents, and the capital costs would be low: Dauvergne told me he could build a ten-megawatt prototype, with a desalination operation, for $50 million. Eventually the company will build 50 MW modules that could be assembled into big central plants or distributed as small modular reactors.

Seventy-four years old, Dauvergne was a gentleman of the old school: he insisted I provide a thorough background, and that he verify my credentials, before he would open up to me. Then he called and said solemnly, "I will help you in every way I can." His Chilean accent, and my inability to find anyone else in the thorium movement who had ever heard of him or his company, gave him an aura of mystery. He used the word "complex" a lot: his background was complex, as was his scientific career. The financial backing for DBI was complex but soon to be complete: "We are already committed for $50 million, in writing," he told me from his office in San Leandro, California. "Now it is a matter of advancing the technology, of bringing people up to speed." Within a week or two, he promised, he would reveal to me the name of the corporation that was putting up the money for the demo plant in northern Chile. As I listened to him describe DBI's glorious future, I thought he was just another thorium believer with a grandiose, vaporous scheme. But it seemed possible he might just pull it off. And I liked him; the force of his vision, like Sorensen's, was irresistible. Thorium seemed to attract such obsessives.

Originally Hector Dauvergne had been a private pilot; that job gave him his passage to Argentina, and later to Uruguay, where he earned a degree in electronics. He made it to California in 1960 and worked where he could: as a farmhand in the Central Valley, on construction sites during California's building boom. He became a U.S. citizen in 1967, in time to get drafted and be sent to Vietnam. He flew resupply missions into the war zone. On his return he studied electrical engineering at the University of California at Berkeley, spending four years there but never earning a degree. ("My education was complex.") At Berkeley he became a protégé of the physicist Franklin Ford, learned of the thorium power experiments going on at Oak Ridge, and immediately conceived his life's mission. In our conversations Hector described this realization as a mystical experience: "I asked this element, I said, 'Tell me what you need to release your energy.' Thorium spoke to me and said, 'I need this and this and this.' I gained a deep understanding of how nuclear power worked and why thorium was so superior to uranium. That conversation went on for maybe a year and a half. That is how I got to the bottom of thorium."

A reactor to burn thorium was his grail, but Dauvergne, the descendant of French colonists in Chile, knew that the only road to it lay through the accumulation of plenty of money. With his brother he founded Dauvergne Brothers in 1965 to commercialize his inventions. He proved to be something of a genius tinkerer: he developed the Dauvergne turbine, an early prototype of a hydrogen-driven automobile engine, for Ford Motors' R&D lab in the 1970s. He started out believing that the future lay in reactors driven by linear accelerators to transform thorium into fissile uranium-233 but later shifted to using a starter fuel—a neutron source such as U-235. In his discussions with Franklin Ford, Dauvergne had reached two key conclusions: a solid fuel thorium reactor should be moderated by graphite and cooled with gas, most probably helium gas. Those premises formed the basis of the Dauvergne thorium reactor.

He spent years fashioning components for what would become the DBI thorium reactor: expansion boilers, a circular heat exchanger, the mock-up core assembly. While thorium R&D came to a halt in U.S. national labs, Hector Dauvergne toiled away, almost alone, like a modern Victor Frankenstein creating a new energy source rather than an artificial human. Among other things, Dauvergne claims that his reactor will power spacecraft. Dauvergne's competitors were conducting research and doing experiments, but "they've never put the reactor together physically," he proclaimed.

Over the years Dauvergne has attracted a number of collaborators, but he never hired an actual workforce, preferring to work in comparative seclusion. He viewed himself as "the Einstein of thorium energy," according to Ken Ricci, a Silicon Valley engineer who worked with Dauvergne for a couple of years.

In his seventies, with enough time and money to pursue his vision, Dauvergne began to go semipublic. Funding was imminent, he assured me. As I finished the manuscript for this book, the world still awaited his ground-breaking announcement.

Thorium was put on Earth for a purpose, Dauvergne assured me. "If there is superior power up above, call it God or whatever, if there was an intelligent creation of the planet, He's not going to leave it with only a bunch of oil that's going to run out, and leave no energy. I believe in an intelligent creator of some sort based upon energy, and thorium is its ultimate creation."

While watching DBI's progress for an actual breaking of ground in Chile, I got to know another septuagenarian who'd committed the remaining years of his life to a revolutionary new form of nuclear power, John Gilleland, the CEO of TerraPower.

TH90 • TH90 • TH90

LIKE DAUVERGNE, GILLELAND HAD SPENT much of his life thinking about sustainable sources of energy. Unlike Dauvergne, Gilleland had plenty of financial backing. In fact, he had the ultimate investor: TerraPower is a spin-off of Intellectual Ventures, the high-tech start-up incubator headed by Nathan Myhrvold, the former chief technology officer of Microsoft. TerraPower is chaired by Bill Gates, a man who could build a couple of nuclear plants and still have a few billion left over to eradicate malaria. "It's amazing how little the world has invested in energy given how big a problem it is now," Myhrvold told the clean-energy website Earth2Tech in 2011. "Gigawatts of power need giga-sized dollars."[3] TerraPower's potential funding reserves were gigasized.

Gilleland's career had been more conventional than Dauvergne's. Educated at Yale and the University of Michigan, he was an accomplished applied physicist who spent many years designing and building large energy systems. The former vice president of energy programs at the engineering giant Bechtel, he was also a futurist: for several years he served as the U.S. head of the International Thermonuclear Experimental Reactor (ITER), the consortium created to lay the scientific and technological groundwork for a fusion reactor. He was accustomed to ambitious, not to say, unattainable, science programs, and he'd been recruited personally by Myhrvold to come out of semiretirement in 2006 to guide TerraPower. "My job was to prove that this is not worth doing," Gilleland told me, "and I'm still here."

This is a fast breeder technology known as a traveling wave reactor. In its original conception, the TerraPower system had an elegance that made it seem almost incredible. In other words, plenty of nuclear engineers didn't think that it would work. In the core the fission reaction, ignited by U-235, would make its way through the U-238 fuel rods, producing highly fissile plutonium-239. The reaction would travel through the fuel like wind rippling across a wheat field, transforming common uranium into rare plutonium until it burned itself out after consuming most of the energy in the fuel. The pool would be cooled by liquid sodium. TerraPower spent years using supercomputers to model the ideal theoretical reactor, and, "lo and behold, what we came up with was a reworking of ideas from the previous century," Gilleland told me.

In fact, the traveling wave concept dates at least to the 1990s, when Edward Teller and colleagues at Lawrence Livermore came up with a reactor design based on a "small nuclear ignitor and a much larger nuclear burnwave-propagating region."[4] The burnwave region, Teller and his associates wrote, "contains natural thorium or (possibly depleted) uranium fuel, and functions on the general principle of fast breeding." The fuel burn-up in such a reactor could reach 50 percent.

Gilleland and Roger Reynolds, the former chief technologist at Areva and now TerraPower's chief technical adviser, wouldn't acknowledge it to me, but the original traveling wave concept had proven too revolutionary to bring to market in a workable time frame, and in 2011 they had modified it significantly. Now the rods would move and the reaction would stay in the same place. Fission would start in U-235 rods in the heart of the core, rather than moving across the fuel, spreading to only the immediately adjacent layer of U-238 rods. As the uranium in this innermost layer transmutes to plutonium and is consumed, the center rods are removed to the periphery and new ones are moved into the fission zone to take their place. TerraPower had designed a beautiful remote-controlled mechanism to rearrange the rods. In theory the reactor could be started and would run for decades without refueling. The amount of unenriched U-238 available could power the world for centuries, until scientists finally work out the kinks in fusion. And TerraPower would license the technology to manufacturers who would sell packaged, prefueled cores to developing nonnuclear countries, eliminating the need for enrichment facilities that could be diverted to weapons programs.

On its face this rod-shuffling machine sounded a lot like the seed-and-blanket design that Lightbridge was supposedly still building, and the word among watchers of advanced reactors was that TerraPower, having found its original concept too expensive, had taken a more conventional route. Indeed the thing about the latest TerraPower design (and, for that matter, the original) was that it was, as Reynolds put it to me, "fuel agnostic." In fact, thorium was not only suitable for the TerraPower machine but also in many ways would be better than uranium. When I first spoke to him in the spring of 2011, Gilleland had just returned from a lengthy stay in India, where he met with the heads of the three-stage program at the Bhabha and Indira Gandhi research centers. TerraPower was looking into the possibility of building thorium reactors based on the traveling wave concept. The technology's main selling point was also its weakness: TerraPower reactors would theoretically require no refueling. The solid fuel rods would have to last 60 years. Given thorium's high burn-up capacity, that would seemingly make it ideal for such a long-lived core. But no fuel element had ever been designed to last that long. TerraPower was struggling to find high-end alloys that would stand up to the powerful neutron bombardment inside a fast reactor core for decades. What's more, Gilleland and his scientists were in love with sophistication and elegance, which is good if you're designing luxury automobiles, not so good if you're building a reactor. In tomorrow's nuclear power industry, efficiency, simplicity, and low cost will matter more than the cleverness and beauty of your design.

Nevertheless, Gilleland expected to have a working prototype before the end of the decade. Like other advanced reactor designers, he does not expect it to be built in the United States. That prospect disturbs him: "I'd love to see the U.S. get off its keister and start innovating again. But there just isn't the drive here that exists in China or India."

TH90 • TH90 • TH90

THE MOST INTRIGUING THORIUM POWER VENTURE I learned about came not from China or India or the United States but from a seemingly unlikely birthplace: South Africa.

A company called South African LFTRs (SAL) appeared on the scene in 2010 and, despite the efforts of its principals to remain in stealth mode, soon captured the attention of the thorium movement. Unlike most of its rivals (with the exception of TerraPower), SAL seemed to be well funded. Like DBI, though, it had a forceful, if somewhat enigmatic, leader.

When I first reached George Langworth, his first question was, "How did you find me?" A former telecom entrepreneur, Langworth was bulkily built, with a buzz cut and a pair of wire-frame spectacles; after seeing a few pictures of him I realized that he bore an odd resemblance to an older Kirk Sorensen. After some hesitation, he agreed to talk to me about his background. Born in Columbus, Georgia, Langworth had lived in New York since the late 1960s and had become a successful telecom entrepreneur and investor: he was among the original backers of the Fiber-Optic Link Around the Globe (FLAG), an early high-speed underwater communications link between the United Kingdom and Japan that passed beneath the Mediterranean, Red Sea, and Indian Ocean, connecting many parts of Europe and Asia. As part of the FLAG project, Langworth helped develop a way of leasing connections on a shared line that were as fast, secure, and reliable as fixed, dedicated lines. In 1997 he spun that technology off into Global Buyers' Collective Network, a privately held provider of data networking services. From there, like others in the early 2000s, he conceived the idea of combining the telecom and power grids into a single system. And the place he decided to do it first was South Africa.

With large and highly developed industries in mining, automobile as-sembly, chemicals, and commercial ship repair, South Africa has by far the largest economy in Africa. Its modern businesses are densely clustered around a few urban centers, in Durban, Cape Town, Port Elizabeth, and the Pretoria-Johannesburg region. These cities are energy starved; blackouts are routine. The government, which restricts how much time people can spend online and how much data they can download, is proposing sharp price hikes for electricity that are expected to cause widespread unrest. Since 2008 the government has embarked on a massive power plant construction program, including at least 20 gigawatts of nuclear power by 2025. It's not clear where the money for these plants is going to come from.

What South Africa does have plenty of is uranium. The twelfth-largest producer of uranium in the world, the country is the scene of a uranium prospecting boom as at least a dozen companies explore the mineral-rich landscape. Enriching that uranium for use in conventional reactors is an expense South Africa can ill afford. Seeing the combination of an energy-strapped economy, a cash-poor government, rich uranium deposits, and a relatively open regulatory environment for foreign corporations, George Langworth sensed an opportunity. And he caught wind of the thorium movement. South Africa would be the perfect place to create a multinational LFTR company. He founded SAL in 2010.

Langworth was secretive and, after our first call, unresponsive. But through sources in the thorium movement, I obtained a leaked copy of the SAL business plan. I emailed Langworth a few times to seek his comment on the plan, but he didn't reply. According to his business plan, he'd assembled an ingenious consortium of businesses, technologists, and investors in South Africa, India, and Germany. The primary investor was Michael "Motty" Sacks, a South African health-care tycoon who'd grown up under apartheid. The machine components would come from Germany, the LFTR chemical manufacturing from India, the reactor design and technology from the United States. By the summer of 2011, in a series of meetings in Oak Ridge, Berkeley, Livermore, and New York, Langworth had enlisted or consulted with many of the leading figures in the thorium movement, including Jess Gehin of ORNL, David LeBlanc, John Kutsch, and Ralph Moir, the former head of conceptual nuclear fusion research at Lawrence Livermore and one of the most influential scientific backers of thorium. Several of these men were already on the payroll. Langworth believed he could build a 100 megawatt LFTR near Cape Town by 2015, with excess heat used for desalinization. He would produce electricity for 25 South African cents (about 4 cents in the United States) and sell it at 65 cents, in a market where the market price is 1.20 rand (100 cents = 1 rand). And he would do it by essentially duplicating the MSR Experiment.

"The initial production goal should be to re-develop a single fluid molten salt reactor that is designed as a 'semi-clone' of the original Molten Salt

Reactor Experiment," Langworth wrote to two SAL associates in April 2011. "This should include a graphite core with channels that the molten salt flows through."

Langworth and his associates planned three preliminary designs: a single fluid with graphite, a single fluid that could be configured to use different fuel scenarios, and a two-fluid design based on the cigar-like "tube-within-a-tube" concept developed by David LeBlanc. To coordinate the scattered R&D teams, Langworth hoped to set up a cooperative agreement with Oak Ridge—the birthplace of the MSR and the home of a number of scientists who would dearly love to spearhead the creation of a new thorium-based nuclear technology—to use the lab's Virtual Environment for Reactor Analysis, originally developed to analyze and evaluate light-water machines, as a distributed model for reactor design.

Building a LFTR in four years would cost something like $3 billion, he calculated. Langworth's vision was as grand as any that Alvin Weinberg or Kirk Sorensen had ever come up with. Ultimately his company would be "an opportunity for a major technology company to build a global private application, [and] operate it as a service that paves the way for mankind's safe, reliable access to a new energy generation technology."

Like most in the thorium power race, Langworth still needed funding. But, so far as I could tell, he had managed to act almost entirely without public knowledge (beyond the core thorium movers) and had assembled the most formidable technical and business team and the most advanced infrastructure to actually build something, once the money started to flow.

What the thorium movement really needed was a patron with a prominent pulpit and access to wealthy supporters. Unexpectedly that person may be a citizen of Ernest Rutherford's country. The latest voice in support of thorium power was an unlikely champion, a Cambridge-educated climate-change activist who'd once worked for the British arm of Friends of the Earth and who had recently been appointed to the House of Lords: Baroness Bryony Worthington.

TH90 • TH90 • TH90

WITH ONE-THIRD OF ITS MEMBERS older than 75 and a sizable percentage of hereditary peers who don't even bother to show up, the House of Lords is usually depicted as a sleepy chamber of dozing dukes ignoring slightly dotty speeches by other members. Whether that was the scene on March 31, 2011, when Bryony Worthington stood to give her maiden speech, the atmosphere changed when she started to talk about the energy crisis. Calling climate change "the political and moral challenge of our time," Worthington noted that "future generations will judge us on how we act on this issue more than any other." Then she mentioned the technology that could solve this challenge: thorium power.

"We need to have a very deep debate about the role of nuclear power," declared Worthington. "The terrible events recently in Japan remind us of the risks inherent in a technology that was primarily developed for Cold War military applications. A civilian nuclear program based on inherently safer, thorium-fueled reactors could engender a paradigm shift in how we view nuclear power."[5]

The youngest member of the House of Lords, Worthington has spent her life campaigning against climate change—in the private sector, for nonprofits, and now as a parliamentarian. After heading up the climate-change program at Friends of the Earth (a far-left group where she has said she felt out of place), she took a sharp detour, going to work as a "policy adviser" (lobbyist) for Scottish and Southern Energy, a major British utility. Then she joined the Labour Government of Tony Blair. She was among the authors of the United Kingdom's 2008 Climate Change Act, which requires the country's total emission of greenhouse gases to be slashed by 80 percent by 2050. In 2008 she left to form Sandbag, a nongovernmental organization formed to enable the public to purchase and retire the emissions credits traded in the European Union's "cap-and-trade" system. Worthington had seen the fledgling emissions-trading market up close, and it was not a pretty sight. In an interview with the *Guardian* newspaper, she compared buying up the credits, otherwise available to corporations to allow them to continue spewing carbon into the atmosphere, to "burning money in front of someone" so they can't spend it on drink and drugs.[6]

Young, passionate, and committed, the baroness stood out among the genteel lords. A *Guardian* columnist called her one of "the smartest thinkers on how best to cut greenhouse gas emissions."[7] She first learned of thorium power when she heard Kirk Sorensen's speech at a 2008 conference, "10 Technologies for the Future."

Not long after her first speech in Parliament, she was approached by John Durham, a businessman and environmentalist. Durham was a friend and occasional financial backer of JoAnne Fishburn, a Canadian documentary filmmaker based in London. A friend in British Columbia had sent my *Wired*[8] story on thorium to Fishburn, who had read and excitedly passed it on to Durham. "I could not believe what I was reading," Durham told me. He proposed forming a U.K. nonprofit to advance thorium power R&D and support the creation of a commercial industry based on thorium plants. Worthington agreed to lend her name and her time, and she proposed a name. She wanted to call it the Weinberg Foundation in honor of Alvin Weinberg.

The foundation made its official debut at an event in Parliament on September 8, 2011. Sorensen attended, as did Richard Weinberg, Alvin's son. The new group quickly generated a strong response from environmental groups in Britain, a country less polarized over global-warming issues than the United States. Fishburn, who had attended the Washington, D.C., conference of the Thorium Energy Alliance just a few days after reading my *Wired* story, was making a documentary on thorium power and agreed to help with publicity. How exactly the Weinberg Foundation would accomplish its mission was unclear, Worthington told me. But its location in London, solid financial backing, and prominent political support promised that the new group would bring attention and support to international thorium efforts that the U.S.-based Thorium Energy Alliance, for all the passion of its members, could not match.

The country of Ernest Rutherford and James Chadwick and the first liquid fuel reactor had some unique qualities that could make it fertile ground for thorium power R&D. For one thing, the British government is deeply committed to reducing carbon emissions and climate change. There is neither the political squabbling nor public skepticism that dogs such efforts in the United States. Paul Madden, a nuclear physicist at Oxford, is one of the world's leading experts on MSR design. And Worthington's status as a member of the House of Lords allows her to convene hearings to examine the technology and the issues surrounding it—something that has not yet happened in the U.S. Congress.

"What we can do is be a good sounding board for making a fuss," Worthington told me. The Weinberg Foundation could well spark a thorium movement in the United Kingdom. But even the baroness acknowledged that the actual building of reactors would most likely happen elsewhere—in China, South Africa, or India. Or even in Japan.

TH90 • TH90 • TH90

I'D MET TAKASHI KAMEI at a couple of thorium conferences and, frankly, had come away with an unclear impression. He seemed bright, cheery (like many Japanese men I'd met, his default expression seemed to be a wide smile), and full of energy. I'd not heard a lot about the prospects for thorium reactors in Japan, and, especially after Fukushima, I discounted them. Japan's history in thorium-based molten salt reactor technology, however, goes back further than any other country, save the United States. It had been driven by a now-elderly scientist named Kazuo Furukawa, Kamei's mentor. Furukawa's interest in MSRs went back to the 1960s when he visited Oak Ridge, ostensibly to learn more about the U.S. fast-breeder program. He left the fast-breeder program and dedicated more than four decades to researching MSRs. In March 2011, not long before his death, he and his son Masaaki Furukawa established a company called Thorium Tech Solutions, Inc. to commercialize thorium-fueled MSRs in Japan. Since the elder Furukawa's death his son and his followers have continued with his work, seeking funds to build a small salt loop experiment to demonstrate MSR technology, with the goal of building a machine to incinerate nuclear waste.

Takashi Kamei, meanwhile, was also trying to keep the thorium flame burning. Kamei's background was similar to others' of his generation. He was born in 1970 and studied nuclear engineering at Kyoto University. Like many scientists in the United States, he found the career prospects in the nuclear power industry dim and, at 29, shifted to the semiconductor industry. In the early 2000s, as global warming became the subject of intense concern in Japan (the scene of the original international agreements on combating climate change, the Kyoto Accords), a former professor asked him to head up a project looking at alternatives to head off catastrophic climate change. As so many other figures in the thorium movement have experienced, a small bell rang in Kamei's mind. "It was very strange," he told me. "The small memory of thorium in my brain returned."

As it turned out, a small group of scientists at Kyoto University had kept alive the study of molten salts—not for nuclear reactors but for fuel cells.

By 2006 Kamei had become a full believer in thorium power. He contacted Furukawa, who encouraged Kamei to pursue the technology by any means possible. With funding from the government's anti–global warming program and from Kawasaki Heavy Industries (one of Japan's major industrial groups), he organized a research team to investigate small MSRs—ten megawatts and even one megawatt. They set a goal of building a prototype by 2015. He even made contact with officials of the Japanese Ministry of Economy, Trade & Industry who, unofficially and privately, supported Kamei's work on thorium power. These discussions went on for three years, "very carefully and secretly." And then he made contact with George Hara.

The founder and chair of DEFTA Partners, a venture capital firm based in Japan and the United States, Hara is one of the major tech investors of the Pacific Rim. A former archeologist in Central America, he got his master's in electrical engineering from Stanford and, after founding DEFTA, became one of Silicon Valley's most prominent venture capitalists and a leading thinker on pervasive ubiquitous communications—the notion that connected computers will be built into just about every object in our daily lives. And, according to Kamei, Hara has become a passionate supporter of thorium reactors. In July 2011, Kamei told me, Hara committed to helping Kamei's group build a prototype LFTR in Japan.

More than any other country in the world, Japan has a complicated history with nuclear power. The Fukushima accident, which occurred 66 years after the destruction of Hiroshima and Nagasaki, cast into doubt the future of nuclear power in Japan—a country with essentially zero fossil fuel resources that gets almost 30 percent of its electricity from nuclear plants. With Toshiba's purchase of a 77 percent stake in the nuclear supplier Westinghouse in 2006, Japan looked to become a leader in next-generation nuclear technology. Fukushima further complicated that already tortured history. If what Kamei told me was true, however, Japan has a chance to lead the world in thorium power technology. The nexus of Japan's universities, government, major manufacturers, and private investors would provide a foundation that, so far, no other country—not even China, which lacks Japan's long history of technological innovation—can match.

In September 2011 Kamei held a series of off-the-record meetings with elected officials and industrialists at the Diet, Japan's legislature. The outcome of those meetings could help determine the future of thorium power in the island nation and the world.

TH90 • TH90 • TH90

IN LATE SUMMER 2011, as I neared the completion of the manuscript for this book, I started getting excited, if somewhat elliptical, emails from Jim Kennedy, the developer and owner of the Pea Ridge Mine. "Bill pending in congress [*sic*]," said one.[9] When I spoke to him, it became clear that it seemed possible that the U.S. Congress might actually do something, not only about the rare earths shortage but also the thorium power future.

Kennedy and John Kutsch, the head of the Thorium Energy Alliance, had spent much of 2011 lobbying on Capitol Hill to get their bill introduced, the one that would create a centralized, federally sanctioned rare earths repository and processing facility. The bill had been drafted but not introduced yet; the bickering about the federal debt ceiling had pushed it back from midsummer to September, when the 112th Congress reconvened. It was to be sponsored by members of both parties—Kutsch and Kennedy declined to name them until a bill actually reached the floor. Kennedy and Kutsch had managed to find a congeries of issues—energy independence, competition with China, exotic materials fueling the domestic high-tech industry, and so on—that Democrats and Republicans could agree on. And the bill would provide for an ongoing supply of funding to fuel a new domestic thorium power industry. It sounded like a commonsense measure. The only problem, of course, was that common sense was growing scarcer in the nation's capital by the day.

A centralized rare earths facility would enable producers to invest together in federally sanctioned processing technology that would separate the supposedly dangerous thorium from the rare earths and preserve the thorium for future energy production. In Kutsch and Kennedy's telling, at stake was not just an important high-tech material and a potential future energy source but the economic well-being of the United States.

The bill would "set up a system where we can sequester thorium and create markets for it, to eventually create energy from thorium, and by so doing to allow any rare earths mine to bring their ore to the central location to be processed, and voila," Kutsch told me. "The Chinese [rare earths] monopoly is dismantled, a new energy source and a new source of exotic materials are created, and the capitalist free-market system is preserved."

He wasn't kidding about preserving capitalism. Kutsch and Kennedy brought an apocalyptic fervor to their double crusade for rare earths and thorium. Almost every one of my conversations with Kennedy during the summer of 2011 tended toward one simple but overarching statement: "If we don't do this, we're fucked."

In chapter 10, I will examine the imperial decline aspects of the race for thorium power. Briefly, Kennedy's argument went like this: the United States has ceded not only its manufacturing industries and its long-term fiscal stability to other countries, most notably China, but also its technological leadership. China already controls the rare earths industry, monopolizing the production of materials crucial to high-tech products as well as the green energy sector; if it's allowed to control the emerging thorium power industry as well, the space for U.S. manufacturing, technology R&D, and energy innovation will be essentially closed. The U.S. economy will continue to decline, energy prices will skyrocket, and the value of the dollar will collapse. General chaos and misery will ensue. That's the Black Swan version.

"This is the last and only opportunity that America will ever have at doing this," Kennedy told me. "If we miss it, we're toast."

Kennedy, of course, has a commercial incentive to paint the bleakest picture possible. It's quite likely that other, unforeseen technologies will emerge in which American engineers and American entrepreneurs will lead the world. The United States still produces many things of value that other countries want to buy. Apple and Google do not need thorium to thrive. And plenty of people in the thorium movement do not agree that yoking a thorium power program to a rare earths bill is the best way to proceed. For that matter, depending on Congress and the White House for any sort of constructive action on far-reaching long-term challenges is probably a fool's game. Kirk Sorensen, for one, has explicitly said he will build a LFTR company without government support. John Kutsch believes, though, that ultimately any successful new nuclear technology must go through the American system: "If you want to sell in the U.S., in Canada, Japan, and Europe, you're gonna have to play by the rules, and unfortunately the folks who still set the rules are Congress and the NRC."

The thorium movement had reached what I called its Trotskyite phase: the revolution was happening, in slow motion perhaps, but it was happening. The groups that had come together to start it, though, were breaking up into opposing camps and competing entities. Consensus was splintering, factions were forming, and the idealism that suffused the *Energy from Thorium* forum three years earlier, when I started my reporting on thorium power, was giving way to a harder-edged pragmatism. Scientists who collaborated with George Langworth (who essentially wanted to build a thorium industry in countries not subject to rigorous international safety and fuel-supply regulations) were accused in some quarters of something close to treason. Free-marketeers sniped at Kennedy and Kutsch's collectivist approach to a rare earths–thorium facility. The few true thorium experts in the West, like David LeBlanc, found themselves hotly pursued by fledgling commercial ventures.

In many ways, this was an encouraging development: for thorium-based reactors to move from theory to reality, money had to be raised and business models had to compete in the marketplace. You can't solve the energy crisis by posting on an online forum. I wondered, though, if the crack-up of the united thorium front, as it were, would doom the movement in the West to endless debate and political posturing. Something had to break, and break soon. In a larger sense, Jim Kennedy was right: the world has only one chance to get this right.

TEN

WHAT WE MUST DO

I n 416 A.D. a Roman bureaucrat named Rutilius Namatianus set out from the imperial capital to his homeland of Gaul. He wrote an account of his journey in verse, *De Reditu Suo* ("Of His Return"), and much of the long poem has survived. Perhaps the most remarkable thing about Rutilius's return was that he traveled by boat up the coast of Italy; in the final decades of an empire famed for its vast network of roads connecting much of the known world, Rutilius deemed a sea voyage safer and quicker: "I have chosen the sea, since roads by land, if on the level, are flooded by rivers; if on higher ground, are beset with rocks. Since Tuscany and since the Aurelian highway, after suffering the outrages of Goths with fire or sword, can no longer control forest with homestead or river with bridge, it is better to entrust my sails to the wayward sea."[1]

The traditional date of the fall of the Roman Empire is 476 A.D., when Germanic mercenaries under Odoacer deposed the last emperor, Romulus Augustulus. Already by the time of Rutilius's homeward voyage, 60 years earlier, decline was inescapable: roads had decayed, aqueducts had collapsed, cities lay in ruins, and harbors had silted. The barbarian invaders who periodically sacked the Eternal City in the fourth and fifth centuries found plenty of evidence of decrepitude: dry aqueducts, fallen monuments, broken water mills. Numerous theories—210 by one count, from poisoning by wine goblets made of lead to moral decadence to the hiring of mercenary armies to replace Roman legionaries—have been put forth to explain the fall of Rome and the onset of the Dark Ages. One of the more recent, first outlined by Joseph Tainter in his 1990 work, *The Collapse of Complex Societies,* has to do with "energy return on investment."[2] Put simply, the energy required to maintain the Roman lifestyle—all those monuments, all those games and spectacles, all those centrally heated bathhouses—became more and more costly as the centuries passed, fertile cropland was depleted, and landscapes were deforested. The empire had to import grain from farther and farther afield, even as the imperial infrastructure—roads, bridges, aqueducts, grain mills, fortifications—fell into disrepair. Remote borders became harder to defend, and the army, the source of Rome's might for a millennium, went underfed and unmotivated. Colonial mercenaries were hired to defend the empire that had conquered their people. At the end, Rome was a hollow shell that crumpled before barbarian hordes.[3]

"The great problem that they faced was . . . they would have to incur very high costs just to maintain the status quo," Tainter told an interviewer for the 2008 documentary *Blind Spot.* "[They had to] invest very high amounts in solving problems that don't yield a net positive return but instead simply allowed them to maintain what they already got. This decreases the net benefit of being a complex society."

A pithier summary of our current dilemma would be hard to formulate. One element of Tainter's thesis, of course, is that no single explanation for the collapse of a sophisticated civilization like ancient Rome would suffice; all 210 theories likely contain shards of truth. But the kernel, as the political scientist Thomas Homer-Dixon describes it, is inarguable: "The Roman empire was locked into a food-based energy system. As the empire expanded and matured; as it exploited, and in some cases exhausted, the Mediterranean region's best cropland and then moved on to cultivate poorer lands; and as its grain supply lines snaked farther and farther from its major cities, it had to work harder and harder to produce each additional ton of grain."[4]

Many historians have noted the curious fact that, while the Romans were the greatest builders and engineers the ancient world ever knew, their technological innovation at some point stalled. Essentially, the development of water-driven grain mills and a vast system of gravity-driven aqueducts notwithstanding, the empire was built and run for a thousand years on the backs of animals and human slaves.

Exhibiting the slightly dotty fascination scholars have with Rome's achievements, Homer-Dixon calculated the energy consumed in building the Colosseum, in terms of the farmland needed to grow the grain to feed the laborers and pack animals. He found that "the Romans had to dedicate, every year for five years, at least 19.8 square kilometers to grow wheat and 35.3 square kilometers—or almost the area of the island of Manhattan."[5] Just to extract, transport, carve, and hoist into place the single keystone required nearly 1,300 square meters (or one-third of an acre) of farmland. The empire exploited the huge reserves of wood, peat, and coal found across its territory, but it was slave labor, fed by grain imported from across the conquered lands, that enabled not only the construction of awesome public buildings and monuments but also the cultivation of a martial aristocracy. When things began to unwind, though, there were not enough free Roman patriots to defend the Eternal City. Imperial Rome fell because it failed to diversify its energy sources.

Remarkably, an innovative technology existed to do just that. Rome was the first civilization to develop all the necessary components for the world's first steam engine—but it never built one for practical use.

In the first century A.D., drawing on earlier works by Cstesibius and Vitruvius, Hero of Alexandria described an *aelopile* (ball of Aeolus, the god of wind), now considered the first device powered by steam. A Greek living under Roman rule, Hero devised a water-filled cauldron heated by fire, with a pair of tubes projecting upward from its lid. The tubes supported a metal sphere, spinning on its horizontal axis, with two nozzles, or tip jets, protruding from it and bent in opposite directions. Expelled through the nozzles, steam generated thrust that spun the ball.[6] It was considered a marvel of ingenuity but something of a parlor trick; though Romans engineers knew how to build cylinder-drive pistons, which are water pumps without return valves, and gearing (as in water mills and clocks), they never thought to use steam to drive machines to perform labor. Why should they, when slaves were plentiful? And so for want of this tantalizingly small imaginative leap, the empire—majestic, built of marble, a thousand years old, and seemingly eternal—collapsed.

Why didn't the Romans figure out how to use Hero's steam engine to replace—or at least supplement—slave and animal labor? From a distance of nearly two millennia, that question echoes the one asked in chapter 6 of this book: Why didn't the United States figure out a way to build and base an industry on molten salt reactors (MSRs)? That question leads, in turn, to the inquiry of this chapter: How can we do so now?

TH90 • TH90 • TH90

IN HER MASTERWORK *The March of Folly,* the historian Barbara Tuchman catalogs a series of critical turning points at which governments and societies, trapped by the status quo and determined to maintain it, in the face of mounting contrary evidence, failed to leap imaginatively to take bold and rational measures to change the course of events. Tuchman called it "pursuit of policy contrary to self-interest."

"Wooden-headedness, the source of self-deception, is a factor that plays a remarkably large role in government," she wrote. "It consists in assessing a situation in terms of preconceived fixed notions while ignoring or rejecting any contrary signs."[7]

However wooden-headed they seem in the light of current events and current science, men like Hyman Rickover and Milton Shaw acted out of unquestionable motives; they opted for light-water reactors and their presumed successors, fast breeder reactors, out of the belief that the evidence and the experience of hundreds hours of experiment and reactor operation so dictated. Nonetheless, Tuchman's definition—"assessing a situation in terms of preconceived fixed notions while ignoring or rejecting any contrary signs"—applies perfectly. Ignoring the potential for thorium power in the 1960s and 1970s was shortsighted. To do so now would be folly, in the same way that the reign of Diocletian—the third-century emperor who prolonged the Roman Empire's death throes by oppressing the peasantry, raising taxes to crushing levels, expanding a stifling bureaucracy, and building up the army through the employment of mercenaries—looks foolish 17 centuries later. Enacting the same policy that hasn't worked to date, only more forcefully, is a concise definition of *folly.* It's also an accurate description of the nuclear industry's strategy for its twenty-first-century renaissance.

We are not Rome (to answer the question posed in the title of a recent best-seller by the journalist and historian Cullen Murphy). But our capacity for folly is equally boundless, while our ability to foresee our own demise far surpasses that of Diocletian or the other rulers of the late empire. Energy policy today, across the West but particularly in the United States, is determined by a toxic blend of wooden-headedness, economic self-interest, scientific ignorance, theology, and technological inertia. The nuclear power industry in particular has been ruled for decades by technological lock-in—the tendency of established technologies to crowd other, competing (and possibly superior) systems out of the market. Perhaps the most prominent and far-reaching example of lock-in is the supremacy of Microsoft and its Windows software platform, a technology considered inferior by many users and analysts that nevertheless controls about 90 percent of the personal computer market. The stagnation of nuclear power technology was recognized by economists long before the industry itself woke up.

"While an appropriate decision at the time, it now seems that light water may have been an unfortunate choice," wrote Robin Cowan, a professor at the University of Strasbourg, in 1990. "One of the interesting features of this history is the belief held by many that light water is not the best technology, either economically or technically."[8]

Overcoming lock-in requires a combination of business incentives, technological innovation, end-user dissatisfaction, and individual determination. With the exception of the last, all are evident in the energy industry today. Whether nuclear power executives and policy makers have the will to carry those forces through to fruition remains to be seen. Nearly every nuclear power executive and expert with ties to the existing nuclear power industry with whom I've spoken in the last three years has uttered some version of what Paul Genoa of the Nuclear Energy Institute said: "You don't just walk away from that and try the new shiny toy, even if the new shiny toy might work better."

In other words, *we can't do it because it's not the way we've always done things.* Given the degree of technological lock-in and institutional inertia that pervades the industry—not to mention the political paralysis that grips Washington—how, then, might the thorium revival actually happen? What must we do?

The baseline condition, the first thing that must happen, is for public perceptions of nuclear power to shift. Put simply, people have to see the risks and rewards of nuclear power more clearly—a task made more difficult than ever in the wake of Fukushima.

TH90 • TH90 • TH90

BY WAY OF COMPARISON, consider two recent energy-related industrial accidents: the natural gas pipeline explosion in San Bruno, California, on September 9, 2010, and the Fukushima-Daiichi nuclear accident that began six months later.

When the pipeline exploded in San Bruno, sending a wall of flame more than a thousand feet into the air, eight people in nearby homes died immediately. Thirty-five houses were leveled and dozens more damaged. Two days later, when the fire was extinguished and the rubble cleared, crews found a crater 167 feet long, 26 feet wide, and 39 feet deep. It was as if an asteroid the size of a refrigerator had gouged the earth.

The San Bruno explosion joined a rash of industrial accidents that included the April 2010 blast that brought down the Deepwater Horizon offshore oil rig, killing 11 workers and spilling about 206 million of gallons of crude into the Gulf of Mexico. Others were the 2006 Sago mine disaster, which killed 12; the collapse of the Kingston Fossil Plant in 2008, which resulted in the largest release of coal ash in U.S. history; and the Upper Big Branch coal mine explosion that killed 29 miners, also in April 2010. Other countries, of course, were not immune: in July 2010 an oil pipeline exploded at the port of Dalian in northeast China, resulting in the worst oil spill in Chinese history.

The San Bruno disaster was also notable in that it involved natural gas, commonly thought of as a cleaner, safer form of fossil fuel than coal or petroleum. It was not unusual, however, in that PG&E, the owner of the pipeline, did its best to avoid blame and legal liability for the accident. Nearly a year after the blast, the company claimed in a court filing that it owed nothing to victims because third-party welders had damaged the "state of the art" pipeline. "The company also indicated it would seek to assign some of the blame for the losses from the explosion to residents themselves," the *San Francisco Chronicle* reported.

In the Fukushima nuclear accident three people died: two young workers trapped in the turbine hall of Reactor 4 (ironically the only unit that contained no fuel at the time of the earthquake and tsunami), and a third man who died at Fukushima Daini, Daiichi's sister plant nearby. The death toll is a facile and sometimes misleading unit of comparison; it will take many decades and billions of dollars to clean up after Fukushima, and the damage to the national psyche of Japan—a country that prided itself on putting to safe and peaceful work the force that had destroyed two of its cities—was incalculable. In the context of the natural disaster of the quake and tsunami, which killed at least 18,000 people, though, the nuclear accident was a footnote. And in comparison with the series of fossil fuel disasters during the previous four years, it hardly rates mention.

"It is the extent of Western civilization—its successes and apparent solidity—that magnifies the shock and terror on the occasions, however contained, of its collapse," wrote the journalist William Langewiesche in his analysis of the 1994 sinking of the ferry *Estonia* in the Baltic Sea, which took the lives of 852 passengers and crew.[9] That description applies powerfully to the Fukushima accident, which engendered a worldwide soul-searching into the history and future of nuclear power and led to at least three nations' forsaking nuclear power altogether. It almost certainly hastened the end of the era of conventional uranium-based reactors.

More than other modern disasters, nuclear accidents inspire shock and terror. Arnold Gunderson, a former nuclear power executive who served as an expert witness in the Three Mile Island investigation and who has become a fierce Cassandra on the subject of nuclear plant safety, called Fukushima "the biggest industrial catastrophe in the history of mankind," an extreme statement regardless of which unit of comparison you use. Several nuclear power advocates I spoke to, meanwhile, referred to Fukushima as "a minor industrial accident," which is probably easier to conclude the farther you live from the east coast of Honshu. The stark differences measure the polarization of attitudes surrounding nuclear power. Either Fukushima proved, beyond doubt, the inherent danger of nuclear power ("If the Japanese can't operate a nuke plant safely, nobody can," commented many analysts), or it proved that even in a natural catastrophe of biblical proportions, far beyond any nuclear plant's design margins, the loss of life and release of radiation from one of the world's largest nuclear power stations was minuscule. There is little middle ground.

TH90 • TH90 • TH90

WHAT FUKUSHIMA ALSO HIGHLIGHTED was the public's misperception of risk. From its earliest days, the nuclear power industry has faced a fundamental risk dichotomy that is hard to clearly explain but has the power to startle the uninformed mind. Simplified, it goes like this: the chances of a significant accident at any single nuclear power plant are very, very small, a claim evidenced by the industry's nearly 60-year record of operation. In many ways, nuclear power is one of the safest industries in the world today. The theoretically possible consequences of a runaway nuclear accident, however, are almost unimaginable. That was the point of the 1957 congressional report that found that the chances of a serious accident were remote but potentially catastrophic. And that's why the Eisenhower administration considered the findings so inflammatory that it did its best to suppress the report.[10]

The media have long reflected this focus on the nuclear threat rather than its relative safety. A study by the University of Pittsburgh physicist Bernard Cohen of coverage in major U.S. newspapers found that, from 1974 to 1978, those papers ran nearly twice as many stories a year on radiation accidents, which caused zero deaths in that period, than on auto accidents (about 50,000 deaths annually). And this was before Three Mile Island.[11]

I'm not arguing that the potential for catastrophic nuclear disasters should be ignored. And the nuclear power industry, by consistently underplaying the risks and being both unforthcoming and untruthful with the public, has compounded its own problems. Nuclear power executives rank with lawyers and politicians near the bottom of the list of trustworthy public figures. They have earned that mistrust. But it's clear that the actual risks of nuclear power, as demonstrated by the industry's record, are far outweighed by the public's fear of radioactivity.

Compare the risks of fossil fuels. Despite the quarrels of some U.S. politicians and a few holdout junk scientists, there's no question that, by continuing to burn huge quantities of coal, oil, and gasoline, modern societies are hastening the onset of disastrous global climate change. The risks are clear: in a century or less, many large coastal cities could be under water, millions of acres of agricultural land will have turned to desert, severe drought will be widespread, resource wars (particularly over increasingly scarce supplies of water and energy) will be common, and so on. The certainty of global warming, however, is for many people abstract and less persuasive than the small but real danger of a nuclear accident that has been amply displayed at Chernobyl and now Fukushima. Even many environmentalists remain implacably opposed to the expansion of nuclear power because their tenuous fears of radioactivity outweigh their certain knowledge of the consequences of continued reliance on fossil fuels.

In part the difference is time. The earthquake and tsunami that devastated parts of Japan happened in a few hours and the ensuing nuclear accidents at Fukushima unfolded over days and weeks. Global climate change is a slow-motion disaster, occurring over decades. Like Romans failing to notice that their bread is getting more costly and their soldiers surlier, we are baking ourselves to death in greenhouse gases while refusing to make the technological leap that would save us.

When it comes to nuclear waste, a subject even more clouded by politics and misinformation, the timescales are inverted: the small risk of radioactivity's leaking from nuclear fuel stored underground is spread out across millennia, while global warming could profoundly change the world that our children and our grandchildren inhabit. "Time transforms risk," wrote the economist Peter Bernstein, "and the nature of risk is shaped by the time horizon; the future is the playing field."[12]

The distortions of time are multiplied by "the false belief that [our] tools could measure uncertainty," wrote Nassim Nicholas Taleb in his best-seller about the miscalculation of risk, *The Black Swan.* "The application of the sciences of uncertainty to real-world problems has had ridiculous effects."[13] Taleb was referring mostly to the financial markets, where the tendency to overlook outlying and unlikely possibilities, which he terms "black swan events," led to the economic crash of 2008. His argument is that, in planning for the future, it's human nature not to account for highly improbable, yet far-reaching, disruptions—the explosion of the space shuttle *Columbia,* the assassination of John F. Kennedy, or the Fukushima tsunami—because our forecasting models are caught in the bell curve of predictability. Nuclear power is one of the few areas of modern life where the opposite is true: the merest possibility of black swan events ("another Chernobyl") has enmeshed the entire industry in a net of fear of the unknown and the unpredictable. (The so-called war against terrorism is another; since 2001 fears of another devastating terror attack have poisoned American life and cost Americans billions of dollars, in ways both visible—long lines at airport security—and invisible—the economic costs of making America less welcome to other nationalities and other creeds.) Despite the nuclear power industry's comparatively clear record of safety, nuclear energy's potential, at least in the United States, has been stifled by risks that are almost too small to measure.

This timorousness is compounded, in the case of thorium power, by an inability to envision, and to make manifest, a different future for nuclear power—one not bound by the cost curves and risk graphs of the uranium era. Risk aversion in the nuclear industry takes two forms: technological and financial. Apart from the comparatively small group of scientists and engineers profiled in this book—many of whom are dismissed as dreamers and dilettantes by the nuclearati—most nuclear technologists are hemmed in by an incrementalism that eliminates the possibility of bold and visionary leaps. Only the smallest and most predictable next step is safe. Only small adjustments to existing technology are acceptable risks. Obvious counterexamples abound in U.S. history; the most obvious are the Manhattan Project and the Apollo program to land men on the moon. Both depended on technologies that did not exist at their inception; both called forth entire new industries based around those technologies, industries that promised no obvious profits before they arose; and both entailed high opportunity costs in that they demanded resources (financial and intellectual) that could, seemingly, have been devoted to other, more obviously attainable objectives. What has changed to render us incapable of summoning the will, confidence, and unity to produce similar achievements against a threat every bit as existential as our opponents in World War II and the Cold War?

The political system, most obviously. Early in the Obama administration, the choice was made to reform the U.S. health-care system rather than produce a far-sighted energy program that would shift the U.S. economy away from fossil fuels and toward sustainable sources, including new forms of nuclear power. Whether that was a wise decision is debatable; unquestionably, though, the current political paralysis in Washington (and, it must be said, the intransigence of the far right) prevents even a rational discussion of such a program, much less a national consensus around a daring and innovative technology such as thorium power.

The other change is in our financial system. Much as the thorium movement would like to see a "new Manhattan Project" for energy, current levels of governmental debt make that impossible, particularly for one based on a technology that most U.S. politicians have never heard of and that a substantial portion of the electorate would reject out of hand. At the same time, the evolution of the private-sector financial system—particularly venture capital and the stock market—favors quick returns over long-term investing, clear "exit strategies" over building new industries and new technologies for the common good, and consumer-focused technologies (mobile phones and applications, social media, new forms of entertainment) over large and complex infrastructure projects. It's a catch-22: thorium power companies have a hard time getting funded without government support, and the government won't support new technologies without demonstrated demand and investment from the private sector.

Boiled down, the challenges facing the thorium movement in the United States are twofold, with the one growing out of the other. First, the nuclear power industry as a whole is hamstrung by objections, mostly based on irrational fears of radioactivity, that nuclear power can never be "truly safe." Second, those objections—along with the broader inertia of an entrenched and stagnant industry—have blocked technological innovation to such an extent that the idea of pursuing a dramatically new system, even one as tested and proven as liquid fluoride thorium reactors, is rejected out of hand. *We can't do it because we've never done it before, and even if we could do it, the public will never support it.* This dual dilemma turns the discussion of LFTRs, and of fourth-generation reactors in general, back on itself in a pointless circle. It always comes back to the same point, and we are trapped in this moment of political paralysis, technological timidity, and financial insufficiency. There must be a way out of this cul de sac.

One way out can be glimpsed by looking to another powerful American industry that risked extinction by a combination of overseas competition, technological obsolescence, and simple folly. At one time it was the largest and most important U.S. industrial sector of all: the auto industry.

TH90 • TH90 • TH90

THE STORY OF THE PHOENIX-LIKE fall and rise of U.S. carmakers is one of the most remarkable business dramas of the first decade of the new century, by any measure a tumultuous ten years for American industry. Here I will quickly review the sequence of events and then draw some big-picture lessons that could be applicable to the energy industry.

In March 2009 President Obama rejected the restructuring plans of GM and Chrysler, saying that their executives had not gone far enough in their strategy for transforming their businesses. At this point, the U.S. auto industry was shedding 120,000 jobs a month, and the two companies were on the brink of complete collapse. Recognizing that the liquidation of a major automaker would have a disastrous effect on the larger economy, which was already in the midst of the most severe downturn since the Great Depression, Obama committed the federal government to what amounted to a takeover of GM and Chrysler, eventually providing a total of $80 billion in loans and investments that gave U.S. taxpayers a large equity stake in the failing carmakers. Both companies entered bankruptcy that spring. While GM reached historic new agreements with its workers and focused on newer, more economical models, Chrysler renewed itself through a merger with the Italian carmaker Fiat.

By the end of 2009 both companies had exited bankruptcy and had begun paying back the loans from the U.S. Treasury. In the first quarter of 2011 both, along with Ford, posted quarterly net profits and were achieving their first sustained net sales gains since the 1980s. By early 2012 GM was once again the world's largest automaker in terms of sales. The industry still faces major challenges—since 2007 more than half of all vehicles sold in the United States have been made overseas, a trend that is likely to continue—but its turnaround has helped stabilize the overall economy and helped the U.S. manufacturing sector add more than 300,000 jobs between the end of 2009 and early 2012. While the Obama administration has been criticized for its inability to bring down total unemployment (which, at this writing, stands at just below 9 percent), critics have been largely silent on the subject of the automakers' turnaround.

That's because the rescue of the automakers went against decades of anti-Keynesian, purist free-market economic doctrine. Declining to let market forces run their course, Obama chose to keep a pair of large companies afloat through large outlays of taxpayer dollars, and he explicitly involved the government in the day-to-day management of the foundering carmakers. In this case, industrial policy—for many years a term of contempt in Washington—worked.

Whatever you think of the principles involved, the bare truth is that the auto industry had become too big, too complex, too globalized, yet deeply intertwined with the fate of the nation for Obama, or any president, to stand by and let it fail. The same is true of the energy industry.

Here I'll acknowledge that the differences in the two industries make this, to some degree, an apples-and-oranges comparison. While the auto industry and the electrical generation sector are roughly the same size in terms of revenues, the power sector is much more heterogeneous, with multiple big vendors (one of which, Westinghouse, is Korean owned), a half-dozen or so major plant operators, and a miscellany of suppliers and service providers. There are only three major U.S. automakers. The vast majority of the vehicle market is retail, consumers buying from dealers; power generation is a mix of wholesale and retail. The power sector is heavily regulated, the auto industry only in regard to vehicle safety and environmental limits. The Big Three are all public companies. The energy industry comprises a hodgepodge of investor-owned utilities, big public corporations, local co-ops, and so on.

Finally, and most obviously, the energy companies are not bankrupt. In some cases they are more prosperous than ever.

What's more, the auto industry's current recovery may well be a deathbed reprieve that only postpones the inevitable, much as Diocletian's draconian reforms stabilized the Roman Empire enough to stave off collapse for another century and a half. But what worked for carmakers could work for nuclear power vendors and producers. From that experience we can draw five lessons for the energy industry:

1. *There must be a sense of crisis.*

   Two major U.S. automakers failing in the midst of a global financial crash constituted a crisis. The majority of Americans see the prospect of catastrophic global warming as a crisis, but so far not enough (at least in Congress) see it that way to push through comprehensive energy policy legislation. At the same time, prominent Republican candidates for president refer routinely to climate-change science as a ploy for research dollars. But the prospect of failing national competitiveness, an unemployment rate of nearly 9 percent, and a staggering trade deficit might be a combination that could generate support for a transformed energy sector, based on alternative technologies, including thorium power, and generating billions in annual exports.

2. *Broad social and national competitiveness goals, not just narrow "shareholder value," must guide the transformation.*

Globalization is an inevitable force, but that doesn't mean that the United States should abandon entire manufacturing and technology sectors to low-cost developing countries. Having a strong and competitive auto industry is seen as a key national interest. So is having a strong, competitive, and innovative energy industry.

3. *New technology must be the basis of the transformation.*

With a few notable exceptions, the nuclear power industry's plans for the next generation can be summarized as "the same, only more so." Any broad nationwide energy strategy should promote, and require, the rapid development and deployment of new forms of nuclear power, especially liquid fluoride thorium reactors. "Generation III" technologies will not solve the energy crisis; thorium power can.

4. *Government support is necessary, but it must be limited and conditional.*

While the big automakers wound up repaying the billions invested in them by the U.S. government, the funding came with a price: they had to replace top executives and invest in new technology and new production systems. Also, the loans were explicitly temporary. Any program to transform the energy sector must come with equivalent conditions.

5. *The transformation must draw on America's competitive advantages.*

Chrysler and GM were able to recover (or begin to recover) because they had strong production and supply-chain infrastructure in place, able managers who were not tethered to the failed strategies of the past, trusted brands, and a deep pool of experienced workers willing (or, rather, forced) to adapt to new terms of employment. The competitive advantages of the U.S. energy sector include the top engineering schools in the world; a vibrant alternative-energy investment market; a pervasive, though aging, power grid; and large numbers of experienced technologists and entrepreneurs who view the energy crisis as the signal challenge of the twenty-first century. Plus, we invented liquid-fuel thorium reactors.

TH90 • TH90 • TH90

THERE IS A PRECEDENT FOR AN INDUSTRIAL POLICY explicitly aimed at fostering specific technologies in specific energy sectors: the lithium-ion battery industry. Since 2009, as part of the American Recovery and Reinvestment Act, the Obama administration has invested about $2.5 billion in stimulus funds in companies, many of them start-ups, making lithium-ion batteries for electric vehicles. Overseen by the Department of Energy's Advanced Research Projects Agency, known as ARPA-E, the program is rooted in a growing consensus that "U.S. corporations, by offshoring so much manufacturing work over the past few decades, have eroded our ability to raise living standards and curtailed the development of new high-technology industries," as a cover story in the *New York Times Magazine* put it.[14]

This is overt industrial policy, but, like the bailout of the automakers, it has elicited few howls of protest from right-wing opponents. The energy policies of India and China—which combine mercantilist economics, the transfer of advanced technology from the West, and a specific focus on next-generation nuclear power, including thorium—raise the question: Is the energy sector too large, too complex, too globalized, and too intertwined with national security to leave to the blind forces of the pure free market? Put another way, are command economies and industrial policy better at building the energy sector of tomorrow than the unguided free market?

In advanced technology sectors like alternative energy, wrote Vaclav Smil, author of *Energy Myths and Reality,* "China increasingly attracts high-tech manufacturing because of its established networks of suppliers and infrastructure—both of which are comparative advantages created through government policies, not granted by nature."[15]

Economic decline is no more foreordained by God than global climate change. And the ideology of unfettered capitalism should not prevent us from thinking and acting strategically about the future of energy and U.S. economic competitiveness. A transformed energy sector based on thorium power could be fueled by intelligent and far-sighted energy policies—admittedly a distant prospect in the current political climate—but the question remains: What would it cost?

The brief answer is more than the $2.5 billion handed to lithium-ion battery makers and less than the $80 billion poured into failing automakers.

Hector Dauvergne, of DBI, believes he can build a solid-fueled thorium reactor for less than $1 billion. Some leaders of the thorium movement have stated that it will take about $5 billion to build the first commercial LFTR, and about $1 billion to build a small prototype. The House of Representatives' draft appropriations bill for all nuclear energy R&D in 2012 totaled $439 million, including $95 million for "nuclear energy enabling technologies." That's not nearly enough, but it's a start.

Several attempts have been made to calculate the cost of electricity from liquid fluoride thorium reactors; these efforts go back to the late 1970s, when a group of Oak Ridge scientists headed by Richard Engel figured that the R&D costs for a thousand-megawatt commercial MSR would be $700 million, or $2.3 billion today, accounting for inflation—in line with current estimates by LFTR start-ups. The overall cost of electricity from a thorium-fueled LFTR plant, however, will depend on several variables, including the cost of capital, length of time to licensing, operating costs such as labor and overtime, the plant's operating efficiency, and so on.

Basing his calculations on the work of Engel's Oak Ridge team and factoring in the estimates for the variables I have just described, Ralph Moir, a scientist formerly associated with Lawrence Livermore National Laboratory, in 2002 calculated the cost of electricity from a LFTR to be 3.8 cents per kilowatt-hour, less than either conventional reactors (4.1 cents/kWh) or coal (4.2). (New natural gas plants are considered to have among the lowest costs per kilowatt-hour, but natural gas is expensive to store and transport, and the gas industry faces serious environmental challenges around the practice of hydraulic fracturing, or "fracking.") Figures published in 2008 by the Nuclear Energy Institute, the industry's trade association, show a much lower cost for conventional nuclear power: less than two cents per kilowatt-hour. That figure is almost certainly low.

Several thorium supporters have calculated the "overnight costs" for a LFTR plant—the actual cost to build and start up a plant, excluding the interest paid to finance the project—and they came up with a range of $2,258 per kilowatt of capacity, plus or minus 30 percent. Thus, a one-megawatt prototype plant would cost $2.2 million in overnight costs, while a commercial thousand-megawatt plant would cost $2.2 billion. Other estimators have come up with costs as low as $1,400 per kilowatt—which, if true, would make the cost of new LFTRs roughly equivalent to that of new natural gas plants. (According to the U.S. Energy Information Administration, the overnight cost per kilowatt for new conventional nuclear plants is $5,335. For conventional natural gas, without carbon-capture and sequestration [CCS] capability, it's less than $1,000; with CCS natural gas is $2,060.)

Much of this is pure speculation. Just getting a dramatically new design through the licensing process of the Nuclear Regulatory Commission (NRC) could take a decade or two and half a billion dollars, which makes building the first LFTRs in the United States an unlikely prospect. For the purposes of this discussion, I will assume that it's possible. At any rate, LFTRs have plenty of characteristics that will make them less expensive to build and operate than conventional nuclear plants—and should make them easier to license, too.

As I described earlier, LFTRs need less plumbing and fewer expensive safety features. Unlike conventional pressurized water reactors, liquid-core reactors do not need massive containment structures or superthick 600-ton pressure vessels (manufactured today only in Japan). Because fuel is reprocessed on the fly, planned downtime is virtually eliminated. While the cost of fuel is a minuscule portion of overall operating costs for today's nuclear power plants, the abundant thorium used in breeder reactors will never see the price spikes hitting uranium purchasers today and will never be in short supply. The simplicity and modular nature of LFTR designs will make them easier to fabricate en masse and assemble on site. And insurance costs for inherently safe reactors will, over time, approach those of conventional power plants burning natural gas, for instance.

There is at least one fly in this punchbowl, and it has to do with starting up a LFTR. You need an external neutron source, specifically a fissile material, to transmute thorium into U-233 and begin the fission reaction. Thorium advocates have almost certainly understated this challenge, but it is hardly insurmountable. The United States has a stockpile of about a ton of U-233, largely left over from the Molten Salt Reactor Experiment. True to form, the government plans to blend this valuable feedstock material with depleted uranium, rendering it useless. (Estimated to cost half a billion dollars, the blending program is set to begin at the end of 2012.) Many in the thorium movement are campaigning to prevent this wooden-headed plan. Once a sustainable number of LFTRs are up and running, they would breed enough fuel to ignite subsequent reactors. Any realistic plan for building and deploying thorium-fueled nuclear plants, though, must account for the cost of fissile feedstock.

Then there are the social costs, which are rarely factored into estimates for either fossil fuel or nuclear power plants. LFTRs are carbon free; their contribution to nuclear proliferation risks is near zero; they not only essentially eliminate the cost of long-term storage of radioactive waste but also provide social benefits by processing existing waste from conventional plants, making it easier to store; and they will jump-start a new era of energy technology innovation that will benefit the companies and the nations that build them in ways impossible to quantify today. Seen in this light, there's no question that thorium power offers the most economical avenue to bring online massive amounts of new generating capacity without adding to current levels of carbon emissions.

TH90 • TH90 • TH90

WHILE A NEW MANHATTAN PROJECT is not going to happen, some form of government support is necessary. Transforming the energy sector is too large a project for the private sector alone. That's the fundamental dilemma that faces the thorium movement. However, there is a middle way, involving higher levels of federal support, a conscious industrial policy to foster advanced nuclear power, and broad incentives to harness the entrepreneurial energy of the private sector.

Congress and the White House should establish a matching funds program, aimed exclusively at two or three technologies, including thorium power, to drive the creation of a Generation IV reactor industry that would swiftly—within this decade—build prototypes and then small commercial versions, first to supplement and later replace the current collection of outmoded plants, then to replace existing coal plants. The government should overhaul the NRC to streamline the licensing process and favor Generation IV designs over incremental, halfhearted advances. It should explicitly benefit start-ups, like TerraPower and Flibe Energy, not just established vendors and manufacturers like GE, and it should promote homegrown technologies like the LFTR. And it should be conditional on not just submitting new designs for licensing but bringing reactors into commercial production in the shortest time possible. With matching investments coming from the private sector, the program should provide at least $2 billion a year and no more than $5 billion, for a total of $4 billion to $10 billion a year.

Many conservatives and liberals alike scoff at the notion of significant funding for new nuclear power—or, indeed, for renewable energy projects such as wind farms and solar arrays. In September 2011 Solyndra, the California-based maker of solar panels, filed for bankruptcy protection after receiving a loan guarantee for more than half a billion dollars from the federal government. Critics of renewables funding, such as Robert Bryce, seized on the Solyndra affair, which threatened to turn into a major political landmine for the Obama administration, as evidence of why the federal government should never "pick winners" in the energy sector.

Here it's important to recall that, as of late 2011, investment by the United States in new energy sources was paltry compared with that of the countries of Western Europe, to say nothing of China. The Solyndra debacle represented less than 3 percent of a loan program that had delivered $19 billion in private capital for reshaping the energy economy, creating thousands of jobs in the worst employment environment since the Great Depression.

For further perspective, keep in mind that, according to the Nobel Prize–winning economist Joseph Stiglitz, in 2007 the Iraq War was costing $720 million *per day.*[16] Big Oil subsidies are also huge in comparison with investment in alternative energy. In 2010 the Government Accountability Office found that the oil industry's waiver for royalties for deep-water drilling in the Gulf of Mexico—originally passed by Congress in 1995, when oil was selling for $18 a barrel—"could cost the Treasury $55 billion or more in lost revenue over the life of the leases." The federal government is already picking winners—it's just backing the wrong horse. Simply requiring big oil companies operating in the Gulf to pay *half* the usual royalties for extracting oil from U.S. territorial waters would fully fund a nuclear power transformation program through 2020, at no cost to U.S. taxpayers. The tobacco industry has funded billions of dollars in health-care and prevention programs to move toward a smoke-free society. Let the fossil fuel industry take a large role in funding the movement toward a carbon-free society based on thorium power.

TH90 • TH90 • TH90

SO, LET US ASSUME THAT A NUCLEAR POWER transformation program is fully funded. The goals are to:

- Build a prototype LFTR within five years

- Commercialize LFTRs starting in 2020
- Bring LFTRs on line at a rate sufficient to replace fossil fuel plants with clean energy sources by 2050

How much power would that be? The United States consumed about 3.8 million gigawatt-hours of electricity in 2010. Coal accounted for 44 percent of that, nuclear for 20 percent. Total U.S. electricity-generating capacity is about 1,000 gigawatts. Under an optimistic scenario for renewable energy production from wind, solar, biomass, geothermal, and so on, let's say that, to reduce carbon emissions enough to stave off catastrophic climate change, by 2050 we must increase the portion of our electricity generated by nuclear power to 50 percent. One half of 1,000 gigawatts is 500 gigawatts, or 500,000 megawatts.

Electricity demand will grow in the next four decades, of course, by as much as 50 to 60 percent in some forecasts. But I'm being optimistic. So let us say that improved conservation technology and changing consumer habits will limit the increase in demand, and we must build enough new nuclear power plants to generate 500 gigawatts by 2050. That's the equivalent of 500 thousand-megawatt nuclear reactors. Between 2020 and 2050 that means building about 17 LFTRs a year. Let's be ambitious and call it 20 new thousand-megawatt thorium plants a year, for a total of 600.

One of the beauties of LFTRs is that they can be mass-produced. Small, modular LFTRs can be built as 250-megawatt machines and assembled into larger plants. Boeing builds about one $200 million jet a day. A modern airliner has many, many more moving parts and greater overall complexity than a 250-megawatt LFTR. If we build, say, four LFTR manufacturing plants a year with each plant producing 20 250-megawatt reactors (five 1,000-megawatt plants) a year (think of the jobs and spillover technological benefits each plant would bring to the state in which it's located), that would just about do it. And from 2050 to 2100 we can build another 400 plants, until we have created 1,000 gigawatts of thorium power. By the end of the century, we will have built a safe, clean energy infrastructure based on a mix of offshore and land-based wind farms, big solar arrays in the West, geothermal, and natural gas plants, layered on top of a baseload power-generating sector of thorium reactors. Particularly in the Southwest, these plants will use excess heat energy to desalinate seawater.

How much will this cost? Technology advances will bring the cost of thorium reactors down rapidly after commercialization, potentially to the cost of a new jet. Call it $1 billion per thousand-megawatt plant. The cost of building 600 thousand-megawatt LFTRs (or twenty-four hundred 250-megawatt machines) would come to $600 billion. Add 15 percent for start-up costs and financing and round up: $700 billion. In comparison, the 2010 budget for the U.S. Department of Defense was $685 billion. In other words, for about what we spend in one year on defense in wartime (which, by the way, is almost as much as all other countries spend on defense combined), we can lay the foundation for a thorium-based, carbon-free energy economy that could last a millennium. And most of that construction cost will be borne by private industry, which, thanks to the expedited licensing and speedy construction of LFTRs, will generate profits from this construction boom in a short timeframe. Consider the costs, direct and indirect, of building any other thousand-megawatt power plant (coal, conventional nuclear, solar, natural gas)—or of doing nothing and allowing climate change to run rampant by midcentury. Building a couple dozen LFTRs a year starts to sound like a bargain.

Alvin Weinberg's vision of a nuclear-powered world running on molten salt reactors will become a reality a couple of generations later than he foresaw.

These are ambitious goals. What, then, must we do to pull them off? To create a thorium energy economy in the next decade, three things must happen at once: funding, licensing reform, and R&D. I have already described the funding mechanism that must be put in place quickly, by the end of 2013. Licensing reform and R&D—including the development and procurement of the needed materials and fuel—must occur in parallel. The president should order the NRC to expedite its licensing process so that the period from application to final approval is no more than five years. That means that by 2015, while a prototype LFTR is being built (at the Savannah River Site, Idaho National Laboratory, or Oak Ridge), companies will begin submitting applications.

At the same time, you must have fuel to start up all those reactors. Two kinds are required: fissile fuel to ignite the chain reaction and transmute thorium into uranium-233, plus the thorium itself. Luckily we have plenty of both. The Department of Energy (DOE) has more than a ton of U-233, produced by past thorium reactor experiments, on hand. Foolishly, the DOE is planning to spend half a billion dollars to blend the U-233 with U-238 and throw it away in the desert. That plan must be scrapped and the U-233 put to good use as starter fuel for LFTRs.

As for thorium, the U.S. Geological Survey estimates that total thorium reserves in the United States are about 440,000 tons, mostly in Montana and Idaho. If we assume that future LFTRs will achieve an energy efficiency of 50 percent (half the available energy in a given unit of thorium is actually converted to electricity), then a single ton of thorium would produce about 12.1 billion kilowatt-hours (or 12.1 million megawatt-hours) of electricity. About 1,650 tons of thorium would satisfy all the electricity needs of the entire world for a single year. Since LFTRs can be run as breeder reactors, producing more fuel than they consume, 440,000 tons is effectively a limit-less supply of nuclear fuel.

TH90 • TH90 • TH90

THE NEXT STEP, once a prototype reactor has been built and tested, is to build a series of liquid fuel reactors to burn up the plutonium and fission products from existing spent uranium fuel. Kirk Sorensen has proposed a type of liquid chloride thorium reactor, a cousin to LFTRs, that will consume transuranic fission products and use plutonium to create uranium-233. The U-233 will be used to start up new LFTRs.

Next we must create the infrastructure to manufacture LFTRs. The expertise to build these machines is dispersed among a cadre of start-ups described in chapter 9, including Flibe Energy, DBI, and so on, as well as among the big nuclear suppliers like GE and Westinghouse, which already, in some cases, have R&D programs for liquid-core reactors. As has happened in the electric vehicle market, the actual manufacturers would likely include established companies (GE), start-ups (Flibe), and joint ventures combin-ing the two. States will compete to host the new plants with tax incentives, university-based R&D support, and training programs to provide the skilled workers. (Here it's worth noting that the Navy has for years been training recruits with only high school educations to be shipboard nuclear engineers. The new thorium power industry will create thousands of skilled, high-pay-ing jobs that do not require a Ph.D. in nuclear physics.)

It does no good to build carbon-free thorium reactors if you don't get rid of the existing nuclear and coal-fired plants. Decommissioning nuclear reactors is a long, involved, and costly process. A typical decom costs $300 million and takes a decade; an extreme case, like the Hanford Weapons Reactor, can cost billions and take many decades. Ways must be found to bring down that cost. One way would be to build new LFTRs on the sites of old nuclear plants and use the new thorium reactors to consume the fission products from the old machines.

As for coal plants, new regulations from the Environmental Protection Agency (EPA) will lead to the retirement of dozens of aging facilities in the next few decades, regardless of what type of new plants come on line. In July 2011 the consulting firm ICF released a report saying that, while shutting down existing coal plants will take longer than foreseen in the EPA deadlines, 30 to 50 gigawatts of coal-fired electricity production will be retired in the coming decade.[17] Total coal-fired generating capacity in the United States is about 314 gigawatts. Shutting down 50 gigawatts of that every decade, and replacing it with safe, clean thorium power, will eliminate coal from the U.S. electrical portfolio by 2070.

These are achievable goals. Remember: the obstacles to creating a thorium power economy in the next 40 years are not technological or even economic. They are political and perceptual. If we don't do it, it will be because we chose not to—not because it was impossible.

TH90 • TH90 • TH90

HERE IS WHERE THE CURRENT nuclear power establishment—the nuclearati—guffaw and roll their eyes. There are a hundred reasons why the scenario I've laid out will not happen, they say. Uranium is inexpensive (for now), the existing reactor population is safe (except when it's not—see Fukushima), plenty of new reactor designs are less radical than LFTRs (which is why they won't make enough of a difference), and so forth. *We can't do it because we've never done it before.*

They are right about one thing: the United States is not likely to be at the center of the thorium power revolution. Here's a more likely scenario.

Discovering the advantages of thorium technology, the Chinese accelerate their program to build a dozen LFTRs in the next 15 years. They recruit the top thorium talent in the world and co-opt the nascent Japanese program, signing lucrative contracts with the top nuclear suppliers in Japan and South Korea, thus compressing further the R&D timeline. By 2030 China is the leading source of LFTR technology—and of raw thorium fuel—in the world.

India, watching its Asian rival move rapidly to the fore in advanced nuclear power, shifts its three-stage program to a more accelerated development schedule based on solid fuel technology from TerraPower and Lightbridge. Using its huge reserves of thorium as leverage with other emerging thorium power nations, such as the United Arab Emirates, India builds a thriving thorium power sector, building reactors at a slower pace than China but, by 2030, becoming a leader in its own right. Enhanced energy security, and the economic power and diplomatic prestige that come with it, allow India to reach a lasting détente with its perennial foe, Pakistan.

Farther east, on the Pacific Rim, both Japan and South Korea rapidly build thorium reactor technology sectors, supplying China and India with the advanced materials and components they need while starting to build thorium reactors of their own. By 2030 the fastest-growing source of electricity in Asia is thorium power; by 2050 liquid fluoride thorium reactors are supplying a significant fraction of the power not only in China, India, Japan, and Korea but also in secondary, technology-importing countries like Vietnam, Taiwan, Singapore, and Indonesia.

Watching this transformation unfold in Asia, the nations of Western Europe—led by France, Norway, and the Czech Republic, already in 2012 the home of significant thorium R&D efforts—belatedly underwrite their own thorium power programs. While the European Union attempts to establish its own thorium power technology sector, low-cost equipment and fuel from Asia prove irresistible, and China becomes the Saudi Arabia of the new nuclear-powered world.

And the United States? Saddled with debt, paralyzed by wooden-headed political opposition to taking action to reverse climate change, and bound to powerful fossil fuel and nuclear power sectors and their well-funded lobbyists, the United States enters an irreversible cycle of declining living standards, diminishing international stature, and ravaged cities. Civil unrest ensues, and the collapse of our political institutions accelerates. Our top graduates, unfulfilled by their professional prospects at home, emigrate to booming technological centers like Shanghai, Singapore, and Seoul. Our vaunted military, unable to procure energy for its far-flung overseas missions, contracts. As in fourth-century Rome, the roads decay, harbors silt up, the legions become disaffected, and the elite retreat into their marble palaces. All because we failed to capitalize on a technology that we once held in our hands.

TH90 • TH90 • TH90

THAT'S A WORST-CASE SCENARIO. And it's hardly inevitable. So what are the chances that Congress will back a technology that, though proven and tested decades ago by American scientists, is seen today as a radical new system? What is the likelihood that the American public will support a new form of nuclear power so soon after Fukushima? How plausible is it that Silicon Valley venture capital funds will provide billions to thorium power start-ups?

One answer to all those questions is: no more likely than it was, in August 1939, when Albert Einstein wrote President Roosevelt to urge development of atomic weapons, that the United States would design, build, test, and detonate a nuclear warhead within six years. The Manhattan Project, which mobilized vast intellectual, material, and technical resources in a short amount of time, is often cited as the paradigm for solving big and complex problems. General Groves's list of essential requirements, born out of his Manhattan Project experience, has become famous in management theory circles: "Put one man in charge, give him absolute authority, keep the chief outside the bureaucracy, use existing government organizations whenever possible, create a small advisory committee," and so on. To that list, based on the experience of the nuclear power industry, I would add, "Keep military concerns separate from economic and energy-related goals." One main lesson of the thorium power debacle is that for too long we have polluted nuclear power policy with rationales and missions produced in the Pentagon. What a disgrace it would be if the United States—the cradle of nuclear physics, the country that first designed and built liquid-fuel thorium reactors, the greatest source of technological innovation the world has ever known—failed to muster the resources and the will to create the energy source for the twenty-first century and beyond.

Forests have been consumed to produce books wondering whether we, as a nation and as a people, are still capable of Manhattan Project–sized achievements and, if not, why not. The declinist school, it must be said, is in ascendance, exemplified most clearly in books like *The End of Influence* by the Berkeley economists J. Bradford DeLong and Stephen Cohen: "The American standard of living will decline relative to the rest of the industrial-ized and industrializing world. . . . The United States will lose power and influence."[18]

My middle-aged, well-educated American friends unquestionably have a waning confidence that they will pass on to their children and their grand-children a world as clean, safe, peaceful, and full of promise as the one we grew up in. Unimaginable budget deficits; rising competition from popu-lous and dynamic Asian countries; declining educational, moral, and cul-tural standards; the rise of seemingly insurmountable environmental crises; the coarsening of public discourse; and the disappearance of inspirational, admirable leadership have all contributed to our sense that we now live in a Spenglerian era of Western decline. A *New York* magazine cover line actually referred to this as the era of "Post-Hope America," the same week *Foreign Policy* magazine's cover headline asked, plaintively, "What Ails America?"

So, when I think about what I've seen reflected in thorium's glossy sur-face in my three years of research, it's simple: hope. Hope that technology can lead us out of the mess into which technology has gotten us. Hope that through divine Providence or intelligent design or the random workings of quantum mechanics, Earth has been granted an inexhaustible energy source that will not destroy the systems and balances that sustain life. Hope that my son, now 12 and a gifted mathematician, may help engineer a thorium power revolution that will solve the energy crisis, dissipate the threat of nuclear an-nihilation, restore a sense of higher purpose and collective endeavor, and keep the lights on for another few millennia at least. In about a century and a half, the Age of Hydrocarbons delivered us a world of shrinking ice caps, resource wars, mass extinctions, and creeping drought. It could take us less than a century to reverse those trends and usher in the Age of Thorium.

For millions of years, thorium has been there, awaiting the right time, the right circumstances, and the right minds to bring it to light and enable it to provide thousands of years of clean, safe, affordable energy. Alvin Wein-berg was right. The time is now. The technology exists, the economics are favorable, and the need is urgent. The choice is ours.

ACKNOWLEDGMENTS

A reporter is only as good as his sources, and in the reporting and writing of this book I had superb sources. Over the last three years, I've conducted nearly 200 interviews on thorium and the future of nuclear power.

First I'd like to thank Richard Weinberg, the son of Alvin, whose email in early 2010 was the real spark for *SuperFuel.*

Then there were the leaders of the thorium power movement, who have dedicated their time and, in some cases, their life's savings to moving this critical energy source forward. Kirk Sorensen, John Kutsch, Kim Johnson, Ralph Moir, Robert Hargraves, David LeBlanc . . . The list is too long to mention them all here, and several of them (particularly Sorensen, Moir, and Julian Kelly) were generous with their time and their expertise in reviewing sections of the book. This book would not exist without them. Though sometimes constrained by their official positions, Jess Gehin and Bruce Patton of Oak Ridge National Laboratory are clear-eyed, dedicated scientists doing important work for the future of energy, and they were extremely generous in helping me understand the basics of nuclear power and the history of thorium reactors. Others who contributed their time in interviews included Hector Dauvergne, Seth Grae, Thomas Graham, John Gilleland, Per Peterson, Charles Hess, Bryony Worthington, and many others. The knowledge and insights of all these experts inform every page of the book, but the errors are all mine.

Adam Rogers, my editor at *Wired,* was the first one to champion the story that would become the book. *Wired* editor in chief Chris Anderson has been both rigorous and encouraging, and I'll be eternally grateful to former colleagues Thomas Goetz and Bob Cohn for bringing me onto the *Wired* masthead in the first place.

My agent, Elizabeth Kaplan, has for years been far more patient with and supportive of my ideas and my proposals than any literary agent should have to be. Forced to wade through pages of unwieldy and overly technical prose, my editor, Luba Ostashevsky at Palgrave Macmillan, firmly yet kindly guided me with the most important of all editorial imperatives: "Simplify." And Polly Kummel is far more than a copy editor: she sharpened the prose and clarified the reasoning on literally every page of this book.

For years my remarkable family, Shawna and Walker, have put up with my absences not only when I was gone but when I was nominally at home. They did the real work of this book. All I had to do was get the words down. My gratitude for their love and support cannot be expressed in words.

Finally I want to mention the men who have done the work of the Hydrocarbon Era. I have seen them in their labors all over the world: the roughnecks of Kazakhstan, the pipefitters of the North Slope, the forklift operators of Baku, the geologists of the Colorado Plateau, the deckhands of the Gulf of Mexico. Their blood literally flows in the fuel that powers our cities and our vehicles. Many, many have died to bring power to the world. This book is also for them.

NOTES

INTRODUCTION

1. See Wernher Von Braun, "Why I Believe in Immortality," in William Nichols, ed., *The Third Book of Words to Live By* (New York: Simon & Schuster, 1962), 119-20, reprinted at ThomasPynchon.com, www.thomaspynchon.com/gravitys-rainbow/extra/von-braun.html.

CHAPTER ONE: THE LOST BOOK OF THORIUM POWER

1. Robert Bryce, *Power Hungry: The Myths of "Green" Energy & the Real Fuels of the Future* (New York: Public Affairs, 2010), 7.
2. Ibid., 261.
3. Ibid., 264.
4. Isaac Sorensen, "History of Isaac Sorensen," MendonUtah.net, http://www.mendonutah.net/history/personal_histories/sorensen_isaac.htm.
5. Joe Bonometti, email message to author, 2009.
6. Ivo Vaša, email message to author.

CHAPTER TWO: THE THUNDER ELEMENT

1. An isotope is simply a specific form of an element, determined by its atomic weight. Thorium-232 and thorium-239 have the same atomic number, 90, but different numbers of neutrons.
2. Susan Quinn, *Marie Curie: A Life* (New York: Simon & Schuster, 1995), 144.

3. Ibid., 144.

4. Marie Curie, *Comptes Rendus.*

5. Quinn, *Marie Curie,* 165.

6. Ibid., 165.

7. David Wilson, *Rutherford: Simple Genius* (Cambridge, Mass.: MIT Press, 1983), 136.

8. Ibid., 137.

9. Ibid., 141.

10. Richard Rhodes, *The Making of the Atomic Bomb* (New York: Simon & Schuster, 1986), 43.

11. Ibid., 153.

12. Ibid., 43.

13. Peter Daniel Smith, *Doomsday Men: The Real Dr. Strangelove and the Dream of the Superweapon* (New York: Macmillan, 2007), 215.

14. Rhodes, *The Making of the Atomic Bomb,* 284.

15. Niels Bohr, "Resonance in Uranium and Thorium Disintegrations and the Phenomenon of Nuclear Fission," *Physical Review* 55 (1939): 418-19.

16. Helen Hawkins, G. Allen Greg, and Gertrud Weiss Szilard, eds., *Toward a Livable World: Leo Szilard and the Crusade for Nuclear Arms Control* (Cambridge, Mass.: MIT Press, 1987), xxix.

17. The cyclotron was named for its inventor, the Berkeley physicist Ernest Lawrence, who won the Nobel Prize in 1939.

18. The story of the Also missions is told in Samuel Goudsmit, *Alsos* (New York: Springer, 1996). Groves' account of the affair is found in Leslie Groves, *Now It Can Be Told: The Story of the Manhattan Project* (Cambridge: Da Capo Press, 1983), 221.

CHAPTER THREE: THE ONLY SAFE REACTOR

1. Most accidents are cataloged on the website for the World Information Service on Energy's Amsterdam-based Uranium Project.

2. Several plants begun in the mid-1970s did not come online until much later; construction on Watts Bar 1, in Tennessee, for example, began in 1972 but the plant did not begin commercial operation until 1996.

3. Philip Macoun, email message to author, 2011.

4. Gerald Grandey, speech to "The Future of Nuclear Energy," symposium, Denver, Colorado.

5. John Byrne and Daniel Rich, *The Politics of Energy Research and Development* (Piscataway, NJ: Transaction Publishers, 1986), 22.

6. "Passive safety" essentially means the plant has an automatic system that responds in case of an accident to avert disaster, such as auxiliary pumps that send water into the core if there is a loss of coolant. "Inherent safety" means that no disaster is possible due to the design of the reactor or the fuel it uses.

7. Many LFTR proponents, including Kirk Sorensen, believe that the best back-end for generating power from a LFTR is what's known as a Brayton cycle, which drives a gas turbine rather than a steam turbine. Either will work, but gas turbines, particularly in combination with secondary exchangers to recycle waste heat, are typically more efficient, converting up to half of the reactor's heat into electricity (some thorium advocates believe that with technological refinement that number can reach 70 percent). A typical steam system in a coal-fired plant is 45 to 49 percent efficient, and the most efficient uranium-fueled lightwater nuclear plants are only 33 to 38 percent efficient.

CHAPTER FOUR: RICKOVER AND WEINBERG

1. Norman Polmar and Thomas B. Allen, *Rickover, Admiral of the Nuclear Fleet* (New York: Simon & Schuster, 1982), 28.

2. Ibid., 75.

3. "The Man in Tempo 3," *Time,* January 11, 1954, www.time.com/time/magazine/article/0,9171,819338-2,00.html.

4. Ibid., 102.

5. "Director Alvin Weinberg: Mr. ORNL," *Oak Ridge National Laboratory Review* 25, nos. 3 and 4 (2002), www.ornl.gov/ornlhome/news_items/news_061020.shtml.

6. Alvin Weinberg, *The First Nuclear Era: The Life & Times of a Technological Fixer* (New York: American Institute of Physics, 1994), 61.

7. Ibid., 4.

8. Einstein wrote Roosevelt to apprise him of the early research on a uranium bomb and the desirability of securing supplies of uranium. Einstein also suggested some government coordination of the research and perhaps some assistance with funding.

9. "Obituary: Professor Eugene Wigner," *The Independent,* January 13, 1995, http://www.independent.co.uk/news/people/obituary-professor eugene-wigner-1567808.html.

10. Richard Rhodes, *The Making of the Atomic Bomb* (New York: Simon & Schuster, 1986), 398.

11. Weinberg, *First Nuclear Era,* 21.

12. Ibid., 25.

13. Winston Spencer Churchill, ed., *Never Give In! The Best of Winston Churchill's Speeches* (New York: Hyperion, 2003), 46.

14. John Keegan, *The Second World War* (New York: Penguin, 2005), 275.

15. Hanson W. Baldwin, "The Atom Bomb and Future War," *Life,* August 20, 1945.

16. Thomas Parrish, *The Submarine: A History* (New York: Viking Adult, 2004).

17. Polmar and Allen, *Rickover,* 111.

18. Weinberg, *First Nuclear Era.*

19. Leland Johnson, *Oak Ridge National Laboratory: The First Fifty Years* (Knoxville: University of Tennessee Press, 1994), 16.

20. Ibid., 26.

21. The Bell X-1 was built by Bell Aircraft, which also built dozens of B-29s, the bombers that dropped atomic bombs on Hiroshima and Nagasaki.

22. Weinberg, *First Nuclear Era,* 38.

23. James Carroll, *House of War: The Pentagon & the Disastrous Rise of American Power* (New York: Houghton Mifflin, 2006), 113.

24. David Alan Rosenberg, "The Origins of Overkill: Nuclear Weapons and American Strategy, 1945-1960," *International Security* 7 (Spring 1983), quoted in Richard Rhodes, *Arsenals of Folly: The Making of the Nuclear Arms Race* (New York: Vintage, 2007).

25. Clark Clifford, "American Relations with the Soviet Union" (also known as the Clifford-Elsey Report), September 24, 1946, Conway Files, Truman Papers, Harry S. Truman Library and Museum (available online).

26. Weinberg, *First Nuclear Era,* 67.

27. Ibid., 69.

28. Ibid., 65.

29. James Carroll, *House of War,* 29.

30. Alice L. Buck, *A History of the Atomic Energy Commission* (Washington, D.C.: U.S. Department of Energy, 1983), 3.

CHAPTER FIVE: THE BIRTH OF NUCLEAR POWER

1. Norman Polmar and Thomas B. Allen, *Rickover, Admiral of the Nuclear Fleet* (New York: Simon & Schuster, 1982), 121.

2. Thomas Parrish, *The Submarine: A History* (New York: Viking Adult, 2004), 435.

3. Norman Polmar and Thomas B. Allen, *Rickover: Father of the Nuclear Navy* (Washington: Potomac Books, 2007), 25.

4. Robert Wallace, "Deluge of Honors for an Exasperating Admiral," *Life,* September 8, 1958, 105.

5. James Mahaffey, *Atomic Awakening: A New Look at the History and Future of Nuclear Power* (New York: Pegasus Books, 2009), 185.

6. Wallace, "Deluge of Honors," 109.

7. Ibid., 110.

8. Alvin Weinberg, *The First Nuclear Era: The Life & Times of a Technological Fixer* (New York: American Institute of Physics, 1994), 43.

9. Weinberg, *First Nuclear Era,* 51.

10. Ibid., 59.

11. Mahaffey, *Atomic Awakening,* 188.

12. Weinberg, *First Nuclear Era,* 59.

13. Ibid., 33.

14. Ibid., 33.

15. "Predicts Atom Will End Limit on Plane Range," *Chicago Tribune,* October 11, 1945, 1.

16. Weinberg, *First Nuclear Era,* 95.

17. Mahaffey, *Atomic Awakening.*

18. Ibid., 284.

19. Weinberg, *First Nuclear Era,* 97.

20. Ibid., 100.

21. H. G. MacPherson, "The Molten Salt Reactor Adventure," *Nuclear Science & Engineering*, no. 90 (1985): 374-80.

22. Todd Tucker, *Atomic America: How a Deadly Explosion and a Feared Admiral Changed the Course of Nuclear History* (New York: Simon & Schuster, 2009), 63.

23. MacPherson, "The Molten Salt Reactor Adventure."

24. Ibid.

25. Alvin Weinberg, "Why Develop Molten Salt Breeders," in ORNL-TM-1851, "Summary of the Objectives, the Design, and a Program of Development of Molten Salt Breeder Reactors" (Oak Ridge National Laboratory, June 12, 1967).

26. Polmar and Allen, *Rickover*, 145.

27. The bill was sponsored by the powerful Chet Holifield, whose role in determining the course of nuclear power in the United States would grow over the next several years, and Senator Albert Gore Sr., father of Vice President Al Gore, now a crusader against global climate change.

28. William Lanouette, "Nuclear Power, 1945-1985," *Wilson Quarterly* 9, no. 5 (Winter 1985), 112.

29. Weinberg, *First Nuclear Era*, 127.

CHAPTER SIX: THE END OF NUCLEAR POWER

1. Alvin Weinberg, letter to W. F. Libby, January 19, 1959, Alvin Weinberg Papers, Oak Ridge Children's Museum, Oak Ridge, Tennessee.

2. "History of Oak Ridge National Laboratory," *Oak Ridge National Laboratory Review* 25, nos. 3 & 4 (2002): chap. 4, www.ornl.gov/info/ornlreview/rev25-34/chapter4.shtml.

3. Johnson and Schaffer, *Oak Ridge National Laboratory*, 97.

4. Alvin Weinberg, *The First Nuclear Era: The Life & Times of a Technological Fixer* (New York: American Institute of Physics, 1994), 108.

5. Ibid., 108.

6. Alvin Weinberg, "Power Breeding as a National Objective," *Nucleonics* 16, no. 8 (1958).

7. Eugene Wigner, letter to W. F. Libby, January 2, 1959. (N.B.: This letter was in the Weinberg papers, which are owned by ORNL.)

8. Wigner to Libby.

9. Ibid.

10. "History of Oak Ridge."

11. James S. Lay, Jr., "Statement Of Policy On Peaceful Uses Of Atomic Energy," March 12, 1955, (Washington, D.C.: Office of the Historian, U.S. Department of State) http://history.state.gov/historicaldocuments/frus1955-57v20/d14.

12. Lanouette, "Atomic Energy, 1945-1985," 110.

13. Weinberg, *First Nuclear Era*, 126.

14. Paul Haubenrceich, J. R. Engel, "Experience with the Molten Salt Reactor Experiment," *Nuclear Applications & Technology* vol. 8, no. 2 (February 1970): 118-137.

15. Charles Barton Jr., "Milton Shaw Part 1," *The Nuclear Green Revolution*, February 21, 2008, http://nucleargreen.blogspot.com/2008/02/milton-shaw-part-i.html.

16. Bill Cabage and Carolyn Crause, "A Chat With Alvin Weinberg," *ORNL Review* vol. 28, no. 1 (Fall 1995).

17. Charles Barton Jr., "Milton Shaw: Part 1." *The Nuclear Green Revolution*, www.thorium1.com/thorium101/history.html.

18. Weinberg, *First Nuclear Era*, 159.

19. Ibid., 162.

20. Ibid., 158.

21. Richard D. Lyons, "Scientists Studying Nuclear Powered Agro-Industrial Complexes to Give Food and Jobs to Millions," *New York Times*, March 10, 1968, 74.

22. John F. Kennedy, "Address at the University of Wyoming," September 25, 1963, The American Presidency Project, http://www.presidency.ucsb.edu/ws/index.php?pid=9433#axzz1i9Kp48Ki.

23. Weinberg, *First Nuclear Era*, 149.

24. "An Evaluation of the Molten Salt Breeder Reactor," Atomic Energy Commission, Division of Reactor Development & Technology, September 1972, 51.

25. "An Evaluation of the Molten Salt Breeder Reactor," 6.

26. Richard M. Nixon, "Special Message to the Congress on Energy Resources," June 4, 1971, University of California Santa Barbara, American Presidency Project, http://www.presidency.ucsb.edu/ws/?pid=3038 #ixzz1ituPN8O9.

27. Bureau of Mines Minerals Yearbook, U.S. Bureau of Mines, 1970, University of Wisconsin Ecology and Natural Resources Collection, http:// digicoll.library.wisc.edu/EcoNatRes/, 1085.

28. R. E. Hollingsworth, letter to Alvin Weinberg, January 26, 1973.

29. H. G. MacPherson, "The Molten Salt Reactor Adventure," *Nuclear Science And Engineering* vol. 90 (1985): 374-380.

30. David LeBlanc, "Too Good to Leave on the Shelf," *Mechanical Engineering,* May 2010, 32.

31. Alvin Weinberg, "Social Institutions and Nuclear Energy," *Science,* July 7, 1972, 33.

32. Cabage and Crause, "A Chat with Alvin Weinberg."

33. Jay E. Hakes, "25th Anniversary of the 1973 Oil Embargo," U.S. Energy Information Administration, September 3, 1998, ftp://ftp.eia.doe .gov/international/25opec/anniversary11_2_98.html.

34. Paul Fine and Holly Fine (producers), "Admiral Rickover," *60 Minutes,* CBS News, December 9, 1984.

35. "The Fugitive Accuser," *Time,* April 8, 1985, www.time.com/time /magazine/article/0,9171,965528,00.html.

36. Weinberg, *First Nuclear Era,* 281.

CHAPTER SEVEN: THE ASIAN NUCLEAR POWER RACE

1. Peter Lavoy, "The Enduring Effects of Atoms for Peace," *Arms Control Today,* December 2003, www.armscontrol.org/act/2003_12/Lavoy.

2. "Nuclear Power in India," World Nuclear Association, June 2011, www. world-nuclear.org/info/inf53.html.

3. Sandeep Dikshit, "Revive R&D in Thorium, Says India," *Hindu,* March 9, 2010, www.thehindu.com/news/national/article221360.ece.

4. Ibid.

5. Anil Kakodkar, "Reversing the Logic of the Nuclear Deal," *The Hindu,* July 3, 2011.

6. M. D. Nalapat, "UPA Sabotages India's Thorium Energy Quest," *The Organiser,* June 26, 2011.

7. Suvrat Raju, and M. V. Ramana, "Strange Love," *Open Magazine,* May 14, 2011, www.openthemagazine.com/article/nation/strange-love.

8. Brahma Chellaney, "Can Corrupt India Handle Nuclear Safety?" *Rediff News,* March 18, 2011, http://www.rediff.com/news/column/india -corruption-nuclear-safety/20110318.htm.

9. M. V. Ramana, "India and Fast Breeder Reactors," *Science & Global Security* 17 (2009): 54-67, www.princeton.edu/sgs/publications/sgs/archive /17-1-Ramana-India-FBR.pdf.

10. Ashwin Kumar and M. V. Ramana, "The Safety Inadequacies of India's Fast Breeder Reactor," *Bulletin of the Atomic Scientists,* July 21, 2009, www.thebulletin.org/web-edition/features/the-safety-inadequacies-of -indias-fast-breeder-reactor.

11. Xu Qimin, "The Future of Nuclear Power Plant Security 'Are Not Picky Eaters,'" *Wen Hui Bao,* January 26, 2011, http://whb.news365.com.cn /yw/201101/t20110126_2944856.htm.

12. Ibid.

13. I wrote about the U.S. rare earth mining company Molycorp in the December 12, 2011 issue of *Fortune.*

14. Cindy Hurst, "China's Rare Earth Elements Industry: What Can the West Learn?" Institute for the Analysis of Global Security, March 2010, Potomac, MD.

15. Hurst, "China's Rare Earth Industry," 13.

16. Lan Lan and Zhang Qi, "China May Appeal WTO Ruling on Resources," *China Daily,* July 7, 2011, www.cs-re.org.cn/en/modules.php ?name=News&file=article&sid=64.

17. Ibid.

18. Karl G. Schneider Jr., "The Rare Earth Crisis—The Supply/Demand Situation for 2010-2015," *Material Matters* 6, no. 2 (2011): 32-37, www.sigmaaldrich.com/etc/medialib/docs/Aldrich/Bulletin/1 /material_matters_v6n2.Par.0001.File.tmp/material_matters_v6n2 .pdf.

19. Malcolm Moore, "Leading Physicist Calls China's Nuclear Programme 'Rash and Unsafe,'" *Telegraph* (UK), June 1, 2011, http://www.telegraph

.co.uk/news/worldnews/asia/china/8549384/Leading-physicist-calls
-Chinas-nuclear-programme-rash-and-unsafe.html.

20. Xiegong Fischer, "China Seeks German Nuclear Know-how," *Deutsche Welle,* June 6, 2011, http://www.dw-world.de/dw/article/0,6542389,00 .html.

CHAPTER EIGHT: NUCLEAR'S NEXT GENERATION

1. Leslie Allen, "If Nuclear Power Has a More Promising Future . . . Seth Grae Wants to Be the One Leading the Charge," *Washington Post,* August 2, 2009, www.washingtonpost.com/wp-dyn/content/article /2009/07/24/AR2009072401847.html.

2. Alvin Weinberg, *The First Nuclear Era: The Life & Times of a Technological Fixer* (New York: American Institute of Physics, 1994), 133.

3. In February 2012 the Nuclear Regulatory Commission announced that it would award a combined construction and operation license to the Southern Co. to two AP1000 reactors at Southern's Vogtle, Georgia, plant. The license was the first issued in the United States since 1978.

4. "Criticality for Fast Reactor," *World Nuclear News,* July 22, 2010, http: //www.world-nuclear-news.org/newsarticle.aspx?id=28097.

5. "Nuclear's Next Generation," *Economist,* December 10, 2009.

6. Weinberg, *First Nuclear Era,* 22.

7. "Report to the Full Commission," Reactor and Fuel Cycle Technology Subcommittee, Blue Ribbon Commission on America's Nuclear Future, June 2011, iv.

8. "Report to the Full Commission," 90.

9. Ibid., 34.

10. Daniel Yergin, *The Prize: The Epic Quest for Oil, Money and Power* (New York: Simon & Schuster, 1991), 24.

11. The dating of the earliest uses of fire is the source of much dispute among archeologists. See, for example, W. Roebroeks, and P. Villa, "On the Earliest Evidence for Habitual Use of Fire in Europe," *Proceedings of the National Academy of Sciences of the United States of America* 108, no. 13 (March 29, 2011): 5209-14.

12. "Plutonium," World Nuclear Association, April 2009, www.world-nuclear.org/info/inf15.html.

13. Kirk Sorensen, "Does 'Conventional' Reprocessing Make Sense?" Energy From Thorium, April 29, 2006, http://energyfromthorium .com/2006/04/29/does-conventional-reprocessing-make-sense/.

14. The four points are adapted from Arjun Makhijani and Michele Boyd, "Thorium Fuel: No Panacea for Nuclear Power," January 2009, paper prepared for the Institute for Energy and Environmental Research and Physicians for Social Responsibility.

CHAPTER NINE: THE BUSINESS CRUSADE

1. Anne Paine, "Different nuke plant fuel proposed," *The Tennessean,* May 8, 2011.

2. The actual number of military reactors includes nuclear-powered naval vessels, which at any given time are in drydock for maintenance, being decommissioned, or otherwise inactive. This number includes eight research reactors operated by the U.S. Army.

3. Katie Fehrenbacher, "From Microsoft to Nuclear, 10 Questions for Nathan Myhrvold," *Earth2Tech,* June 17, 2010, http://gigaom.com/clean tech/from-microsoft-to-nuclear-10-questions-for-nathan-myhrvold/.

4. E. Teller, M. Ishikawa, L. Wood, R. Hyde, and J. Nuckolls, "Completely Automated Nuclear Reactors for a Long-term Operation II: Toward a Concept-Level Point-Design of a High-Temperature, Gas-Cooled Central Power Station System," paper presented at the plenary session of the 1996 International Conference on Emerging Nuclear Energy Systems, Obninsk, Russian Federation, June 24-28, 1996.

5. "Baroness Worthington's House of Lords maiden speech," March 31, 2011, online video, Vimeo, http://vimeo.com/21912309.

6. Leo Hickman, "Sandbagged: Dealing a blow to carbon trading," *Guardian* (UK), September 12, 2008, http://www.guardian.co.uk/environment /2008/sep/12/carbonemissions.carbonoffsetprojects.

7. Ibid.

8. Richard Martin, "The New Green Nuke," *Wired,* December 2009, http://www.wired.com/magazine/2009/12/ff_new_nukes/.

9. Jim Kennedy, email to the author, July 29, 2011.

CHAPTER TEN: WHAT WE MUST DO

1. Rutilius Namatianus, *De Reditu Suo,* translated by J. Wight Duff and Arnold M. Duff, *Minor Latin Poets,* Loeb Classical Library, vol. II, 753-829.

2. Joseph Tainter, *The Collapse of Complex Societies* (Cambridge: Cambridge University Press, 1990).

3. The general thrust of the first sections of this chapter, and some of the historical information, come from a talk given by Ugo Bardi at the "Peak Summit" in Alcatraz, Italy, on June 27, 2009, entitled "Peak Civilization." The talk was transcribed by Bardi and posted online on July 22, 2009, by Energy Bulletin: http://www.energybulletin.net/node/50025.

4. Thomas Homer-Dixon, *The Upside of Down: Catastrophe, Creativity, and the Renewal of Civilization* (Washington, D.C.: Island Press, 2006).

5. Ibid.

6. *The Pneumatics of Hero of Alexandria,* trans. Bennet Woodcroft (London: Taylor Walton & Maberly, 1851), http://www.history.rochester.edu/steam/hero/.

7. Barbara Tuchman, *The March of Folly* (New York: Ballantine, 1984), 7.

8. Robin Cowan, "Nuclear Power Reactors: A Study in Technological Lock-in," *Journal of Economic History* 50, no. 3 (September 1990).

9. William Langewiesche, *The Outlaw Sea* (New York: North Point Press, 2004), 101.

10. William Lanouette, "Nuclear Power in America," *Wilson Quarterly* vol. 9, no. 5 (Winter 1985): 110.

11. Ibid., 106.

12. Peter Bernstein, *Against the Gods: The Remarkable Story of Risk* (New York: John Wiley & Sons, 1996), 15.

13. Nassim Nicholas Taleb, *The Black Swan: The Impact of the Highly Improbable* (New York: Random House, 2010), xxii.

14. Jon Gertner, "Does America Need Manufacturing?" *New York Times Magazine,* August 24, 2011.

15. Vaclav Smil, "The Manufacturing of Decline," *Breakthrough Journal*, July 2011.

16. Kari Lydersen, "War Costing $720 Million Each Day, Group Says," *Washington Post*, September 21, 2007.

17. "ICF International Integrated Energy Outlook Predicts Up to 40 GW of Coal Plant Retirements in the Next Two Decades," company press release, October 10, 2011, http://www.icfi.com/news/2011/third-quarter-2011-integrated-energy-outlook.

18. Bradford DeLong and Stephen Cohen, *The End of Influence: What Happens When Other Countries Have the Money* (New York: Basic Books, 2010).

INDEX